Who Rules the Waves?

WHO RULES THE WAVES?

Piracy, Overfishing and Mining the Oceans

Denise Russell

PlutoPress
www.plutobooks.com

First published 2010 by Pluto Press
345 Archway Road, London N6 5AA and
175 Fifth Avenue, New York, NY 10010

www.plutobooks.com

Distributed in the United States of America exclusively by
Palgrave Macmillan, a division of St. Martin's Press LLC,
175 Fifth Avenue, New York, NY 10010

British Library Cataloguing in Publication Data
A catalogue record for this book is available from the British Library

ISBN   978 0 7453 3005 1   Hardback
ISBN   978 0 7453 3004 4   Paperback

Library of Congress Cataloging in Publication Data applied for

This book is printed on paper suitable for recycling and made from fully managed
and sustained forest sources. Logging, pulping and manufacturing processes are
expected to conform to the environmental standards of the country of origin.

10   9   8   7   6   5   4   3   2   1

Designed and produced for Pluto Press by
Chase Publishing Services Ltd, 33 Livonia Road, Sidmouth, EX10 9JB, England
Typeset from disk by Stanford DTP Services, Northampton, England
Printed and bound in the European Union by
CPI Antony Rowe, Chippenham and Eastbourne

For Mingaloo

Contents

List of Figures and Tables   ix
Acknowledgements   x

Introduction   1

1 **Freedom of the Seas**   6
   Early attempts to close off the seas   6
   Pirates, privateers and the domination of the seas   7
   Grotius' arguments for freedom of the seas   14
   Replies to Grotius defending closure of the seas   17
   Limits to freedom of the seas   21
   Grotius' principles in the current law of the sea   23
   Climate change, rising sea levels and the displacement
      of island communities   26

2 **Underwater Non-living Resources**   29
   Who has a claim?   29
   Antarctica and the Southern Ocean   32
   The Arctic Ocean   33
   Ecological threats from oil and gas activities in
      the Arctic   35
   Stresses on the Arctic from climate change   40
   Ocean acidification   43
   Different ways of valuing the polar regions   45

3 **Underwater Cultural Heritage**   47
   What is underwater cultural heritage?   47
   Salvage Laws   49
   Treasure salvors and ownership   51
   National ownership   52
   Common heritage   55

4 **Modern Piracy and Terrorism on the Sea**   60
   The *Alondra Rainbow*   60
   The law of the sea and contemporary piracy   61
   Why piracy now?   67
   The rise of piracy in Somalia   70
   Pirate attacks on private boats   75
   Terrorism on the sea   76

5   **The Fishing Wars**   84
    The cod wars   84
    The turbot war   88
    Fish piracy   92
    Threats to fish populations from climate change and
        ocean acidification   100
    The war on fish   100

6   **Cetaceans and the Sea**   105
    Whales and dolphins   105
    Cetaceans and morality   107
    Threats facing cetaceans   108
    Protection agencies   117

7   **Sea Gypsies**   121
    Sea gypsies: people without an address or 'names that
        can be found in books'   121
    The sea as home   128
    Threats to sea-gypsy cultures   132
    Sea borders, shark fishing and cultural survival   134

8   **Indigenous Sea Claims**   137
    Ownership as belonging   137
    Contemporary attempts to assert ownership of the
        oceans by indigenous groups   140
    Australian High Court decisions on Sea Rights   142
    Indigenous sea rights and environmental threats   148

9   **Protection of the Oceans**   150
    Ownership of coastal areas   150
    Ownership of international waters   153
    International ocean governance   158
    Implementation of a new ocean management regime   163

Notes   165
Index   186

List of Figures and Tables

FIGURES

1.1 The Mediterranean in the sixteenth century
       including key centres of piracy and privateering   8
1.2 The British Seas in 1635   20
6.1 Migration of humpback whales   106
7.1 Moken family going about daily tasks in their boat   123
7.2 Inside a Moken boat   124
7.3 Moken children playing   132

TABLES

2.1 Effects of Petroleum or Individual Polycyclic
       Aromatic Hydrocarbons on Organisms   38
4.1 Piracy and Armed Robbery Against Ships:
       Actual and Attempted Attacks Since 1994   64
4.2 Actual and Attempted Hijacks of Ships Since 1994   65

Acknowledgements

I would like to thank David Castle from Pluto Press for his wise editorial advice, and the anonymous reviewers for their thoughtful comments. Hal Pratt gave me tireless support and reacted appropriately to outlandish speculation. Tony Miller, Graham Staines, Gerd Weidemann, Helen Wilson and Paul Sharrad kindly read drafts, provided me with critical feedback and thereafter alerted me to relevant activities and sources. Frances Patterson ably assisted me with the index. San MacColl and Truda Gray also led me to some relevant materials. Jane Lymer helped me to bring the book to completion with her invaluable clerical assistance. Diana Wood-Conroy and Dorothy Jones kept me looking east. Mingaloo, an albino humpback whale, and other sea creatures, force me to care about their habitat.

The University of Wollongong in Australia is very generous in supporting my research under their fellowship programme.

Introduction

August 2007: the mini-submarine *Mir-I* disengages from a Russian research vessel and Artur Chilingarov powers the vessel down to the bottom of the Arctic Ocean near the North Pole. *Mir-I* lands smoothly on the yellow seabed. The purpose of the visit was partly to collect fauna and flora, but they were not in evidence. The prime motive, however, was to place a flag made out of titanium that would last for centuries, in order to stake a claim on the seabed for Russia. The tricky part was how to get back from there, 4,200 metres under the sea. A gap had been cut in the ice above but if it wasn't located the submarine could be trapped.

The opening was found and Sergei Balyasnikov, a joyous spokesman for the Russian Arctic and Antarctic Institute proclaimed: 'This may sound grandiloquent but for me this is like placing a flag on the moon, this is really a massive scientific achievement.'[1] It was certainly brave and theatrically captivating but there was no resounding scientific result. In 40 years time we won't be reliving the moment. Rather, the move was part of a political game to assert ownership over the seabed of the Arctic Ocean for the purposes of oil and gas exploration. The other players entered on cue. The Canadian Foreign Minister Peter MacKay said: 'This isn't the fifteenth century. You can't go around the world and just plant flags and say "We're claiming this territory."'[2] Tom Casey from the US State Department next entered the stage with the lines: 'I'm not sure of whether they've put a metal flag, a rubber flag or a bed sheet on the ocean floor. Either way it doesn't have any legal standing.'[3] The play continues with all the Arctic Rim countries taking up positions and the European Union announcing plans to build a ship capable of boring up to 1,000 metres into the seafloor in water depths of 5,000 metres under the ice. The ship will further the scientific research of the Arctic Ocean seabed begun by the Russians. The anticipated completion date is 2014.

If Russia wants to stake a claim under the Arctic Ocean and mine there, should anybody care? One of the contentions of this book is that everybody should care. It is not just that there are conflicting legal claims to rights over this seabed, it is the fact that the Arctic is one of the last fragile wilderness areas on earth and mining under the Arctic Ocean poses unacceptable risks to that land/sea environment. Such mining would have an intimate link with climate change, inevitably increasing greenhouse warming with a global reach.

In a few decades we have moved from thinking of the oceans as places of wildness and robustness to an understanding that what we do in them can determine the state of the planet. We may now have the tools to mine under the ice, but quite possibly not the wisdom to proceed with caution. As recently as 1955 Rachel Carson completed her trilogy of books on the sea with a chapter called 'The Enduring Sea' and the comment: 'And so we come to perceive life as a force as tangible as any physical realities of the sea, a force strong and purposeful, as incapable of being crushed or diverted from its ends as the rising tide.'[4] Such optimism is now out of place, but it was sentiments such as these that lay behind the idea that the oceans should be areas of freedom where people could do whatever they wished: fish, dump, pollute, navigate, somehow the ocean would cope, though Carson would have deplored this implication. We now see reports that paint a worrying picture of the future of fish populations, of whales and of all the small creatures with calcium carbonate skeletons vulnerable to ocean acidification, creatures that play a vital role in the marine food web.

So do we keep going on the same track, trying to reprimand nations for overfishing or whaling with the regulatory regimes now in place, or should we be looking for something new? There is another threat in the seas and that is the rise of piracy. This has no apparent link with the environmental pressures on the oceans. It is, however, of vital importance to the lives and livelihoods of many users of the sea. The organisations and policies we now have in place for the oceans are failing us. The fact that what lies on, in and beneath the oceans is under attack leads to international unrest and conflict. The United Nations Law of the Sea, while standing behind the freedom of most of the high seas, is encouraging greater 'privatisation' of the sea by coastal states likely to govern the use of living and non-living resources in those areas for the national benefit. However, it is the international community that will be the winner or loser in the fate of the oceans. So there needs to be a more effective way of allowing that community to exercise control over the oceans. Neither the free seas regime nor the creeping national jurisdiction are effective ways of allowing the international community to exercise control over the oceans.

How did we get to the position where the notion of the freedom of the high seas is built into international law? In Chapter 1 I explore the contemporary carving up of the oceans arising partly from the consequences of piracy. The influence of Hugo Grotius, a Dutch lawyer writing 400 years ago, is highlighted, as it is his thinking that lies behind the key concepts in the Law of the Sea.

In Chapter 2 I return to the Russian flag. Territorial disputes arose in 2007 over the Arctic Ocean and the resources on the sea floor among Russia, Denmark, Norway, Canada and the United States. This seabed contains huge oil and gas reserves. As the ice melts with climate change, extracting those reserves may become feasible. After Russia placed the flag under the North Pole, Denmark (through Greenland) countered by claiming the area as Danish territory. Canada followed and revealed plans to establish an Arctic port. The US is interested in extending its territory from Alaska into the Arctic. Research is being conducted to try to validate these various claims, which have the potential to generate international tension. In 2008 the European Union announced its intention to be involved in Arctic resource exploration. Canada in particular viewed this announcement with suspicion. In 2009 Argentina and Britain put in competing claims for the continental shelf around the Falkland Islands, South Georgia and South Sandwich Islands.

With advances in technology used to explore the seabed, it is now possible to discover valuable heritage under the seas in areas that were not previously accessible. In 2007 tension arose between some private treasure hunters from the US and the Spanish Government over the discovery of a wreck in the Atlantic Ocean containing $747,000,000 worth of silver and gold Spanish coins. Although the treasure had been off loaded and stored in Florida, the Spanish Government forcibly impounded one of the treasure hunters' ships in the port of Algeciras on the grounds that the coins belong to Spain. In addition, the Spanish argued that the coins had been transported with the complicity of the British. This is a foretaste of battles to come not only over shipwrecks but also concerning other items of submerged cultural heritage. These issues are taken up in Chapter 3.

Piracy and terrorism on the sea are discussed in Chapter 4. Modern pirates threaten commercial ships with theft, hijacking and ransom demands in the Gulf of Aden, Somalia, Nigeria, Indonesia, Tanzania, Bangladesh and Vietnam. Political terrorism on the sea first drew international attention when a passenger liner was hijacked in 1985 and it continues to be a concern.

The global depletion of fish populations has the appearance of a war on fish. With high-speed boats able to use new technology to accurately locate schools and efficient equipment to haul in catches, the fish are very easy targets. The general unwillingness to stick to conservation measures makes humans complicit in the possible extermination of fish. The conflict between the desire to protect fish populations and the desire to hunt them beyond sustainable limits has generated fishing wars between Britain and Iceland and between Canada and the EU. Some of the 'fish pirates' who take fish in violation of international agreements in the Southern Ocean have been fined large amounts in French and Australian courts. The complex web of fishing problems is presented in Chapter 5.

Chapter 6 deals with cetaceans. Whales and dolphins have wide appeal and do enjoy some protection under the International Whaling Commission. However Japan, through the loophole of scientific whaling, continues to provoke widespread condemnation for killing many whale species in the Southern Ocean even in a designated marine sanctuary. Opposition is escalating especially from Australia and the US. Less well known is the threat to whales and dolphins from habitat degradation including marine pollution. Should humans take a moral stand here and act to limit the death of cetaceans and the destruction of their habitat?

One way of thinking about these problems with the oceans is to see them in terms of ownership. The conflicts over the Arctic and underwater cultural heritage are about ownership of the seabed. Contemporary pirates thrive in waters owned by coastal states unable or unwilling to offer effective resistance. Once pirates are on the high seas, unowned territory, there has been little interest in their pursuit until the extent of Somali piracy in 2008–9 forced action from the international community. Humans can wage a war on fish because it is commonly believed that fish have no moral worth and no rights in any sense to ownership of their habitat. Nations dispute the rights of other nations to lay claims to territory for their exclusive fishing activities. Defenders of whales and dolphins see these creatures as having a moral right to their habitat, the ocean.

Some indigenous groups in Southeast Asia, Australia and New Zealand contest the ownership of the inshore areas of the ocean by coastal states. These groups have historic links with the sea stretching back hundreds or, in the case of the Australian Aboriginals, thousands of years. The sea gypsies of Southeast Asia live on the sea. In the twenty-first century their cultures are in jeopardy mainly because they have no ownership rights to their home, the sea territory, and so can be forced out of it. These unusual cultures are described in Chapter 7. Coastal indigenous populations often do not make a sharp distinction between land and sea. While they have been accorded some land rights the courts have been reluctant to grant ownership to any waters. The New Zealand Treaty of Waitangi did accord some sea rights to the Maori people but these have been steadily eroded. A recent High Court decision in Australia marks a significant step to granting sea rights for certain indigenous groups in northern Australian and it has the potential for wider applicability. Chapter 8 focuses on this issue of indigenous claims to sea rights.

A formidable force involved in the fate of the oceans favours a largely unregulated sea. This is the group of corporations that make use of the oceans in diverse ways: the shipping companies, exporters (especially from the automobile industry), commercial fishers and whalers, and marine heritage explorers. Freedom to make use of the sea is assumed, and pollution is given far too little attention. Tighter regulation is resisted. Scientific objections to how we use the land and the sea arising from research on climate change and ocean acidification must, however, be heard. We cannot maintain present-day practices on land and sea and expect to have a healthy ocean in the near future.

Climate change issues are explored throughout the book, interwoven with the discussion of sea refugees, undersea mining, fishing and whaling. In the final chapter I outline a new schema for ocean governance that extends the reach of *international* control and summarise the reasons why I think we should be moving in that direction. My contention is that the Law of the Sea is now part of the problem with the oceans and a radical reorganisation of ocean ownership is needed. Instead of a free-for-all, the high seas should be owned by the international community and regulated to ensure equity between nations and generations. The need for this reorganisation emerges from changed material conditions in the twenty-first century and the realisation that the old ways of ruling, or not ruling, the waves are severely limited in their ability to ensure the safety of travel on the seas or the fair distribution of resources, or to protect the oceans and their life forms and, as a consequence, life on earth.

1
Freedom of the Seas

If each party knew what the other should do, then conflict would be unlikely. And this worked at every level, from the most minor transaction between two people to dealings between nations. International law, after all, was simply a system of manners writ large.

Alexander McCall Smith, *The Sunday Philosophy Club*[1]

Debates over whether the seas should be free for all to navigate and exploit or closed for the benefit of particular states have formed an important strand in western thinking about the oceans for at least 2000 years. The current position, to be explored below, is that the high seas – the areas beyond a 200 nautical mile limit from coastal states – are free for anyone to navigate or exploit while abiding by some fishing regulations that have proved very difficult to enforce. Sovereign rights over the seabed may extend further out. The contemporary concern about the Arctic Ocean relates to this point and is taken up in detail in the next chapter.

EARLY ATTEMPTS TO CLOSE OFF THE SEAS

When coastal states first attempted to close off surrounding seas for their exclusive use this often simply meant that they were able and willing to defend the sea territory against foreigners. At other times their claims had legal backing within the law of the coastal state. Foreign states often objected, though not always, and it was sometimes seen as a duty of sovereigns to safeguard the neighbouring sea.[2] Venice laid claim to the whole of the Adriatic Sea in the thirteenth century. Levies were imposed on foreign shipping. Navigation by foreign vessels was at times prohibited. Venice's sovereignty over the sea arose by force but was recognised by some other European powers as well as the Pope. Venice's position benefited some other states by keeping out piracy in the Adriatic.[3] For many centuries Venice's claims were celebrated in a yearly festival called 'espousing' the Adriatic:

On Ascension Day the Doge was rowed to the strains of music in a magnificent gilded state barge, the *Bucentaur*, to the channel of Lido, where he caste a ring into the water, exclaiming as he did so, 'We espouse thee, O Sea, in sign of a real and perpetual dominion.' The Papal nuncio and representatives of other states assisted at the ceremony.[4]

The development of an international Law of the Sea began with Hugo Grotius, but the background to his theorising grew out of the problem of piracy, and did not go in the direction of tighter regulation as one might expect. Piracy came into these matters in two respects. As the story unfolds below we will see how Spain and Portugal attempted to close off the Atlantic in part to deter the North African pirates coming after their shipping. These pirates had achieved control of the Mediterranean Sea in the sixteenth century and seemed unstoppable. Of course economic incentives to close the seas also existed. It was deemed illegal for non-Portuguese vessels to transit the Atlantic Ocean and violations led to the Portuguese treating foreign crews as pirates. Foreign trading ships were 'legitimately' attacked. This was followed by objection and retaliation from their trading rivals, especially the Dutch. Grotius' was employed to provide arguments concerning the right of the Dutch to travel out to the East Indies. These arguments took the form of a general defence of the freedom of the seas. In a curious replay we may now see the Law of the Sea being re-visited in response to the activities of contemporary pirates, but this time it is the freedom of the sea that will be questioned.

PIRATES, PRIVATEERS AND THE DOMINATION OF THE SEAS

Waves of piracy dominated the Mediterranean Sea from at least 1,000 years BC. Approximately 100 years before the international law of the sea began to be seriously discussed, an event occurred which set in train piracy activities that were to have an important influence on how the law developed. After the fall of Granada to the Christians in 1492, its Moorish inhabitants were forced either to convert to Christianity or go into exile.[5] Several hundred thousand Moors left the Iberian Peninsula crossing the Strait of Gibraltar and arriving in northern Africa looking for ways to make a living in this poor place.[6] The seed for a Holy War waged by Muslims against the Spanish and other Christians was sown.[7] These displaced people built large, fast rowing ships for pirate raids. Just twelve years after this exile began two Papal galleys were attacked by Moors in a clever manoeuvre. After overpowering the first galley, the pirates changed clothes with their captives and towed the pirate ship thus fooling the second galley into coming closer. When it did, the pirates successfully boarded and overpowered it too. This raid was carried out by Arouj Barbarossa who, alongside his brother Kheir-ed-din, led the pirates from the Barbary Coast of North Africa.[8] From this time the Mediterranean again began to be ruled by pirates. Figure 1.1 shows the key centres.

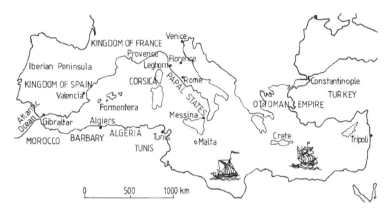

Figure 1.1   The Mediterranean in the sixteenth century including key centres of piracy and privateering. Map compiled by Hal Pratt.

Kheir-ed-din, also known simply as Barbarossa, organised a strong team including the '"Jew from Smyrna" who was suspected of black magic because he could take a reading of a position at sea by means of the crossbow'.[9] They took control of the Mediterranean Sea, sometimes with the help of rulers along the coasts to whom the pirates paid protection money. In return the coastal states would offer safe haven and also markets for their wares, captured people or stolen goods. When Barbarossa's deputy, Hassan, took control of Algiers, the slave barracks were so full that it was 'commonly said that a Christian was scarcely fair exchange for an onion'.[10] One of Barbarossa's powerful allies was a woman, Sida Al Hurra, a regent of the western coast of Morocco who led pirate raids against Spanish and Portuguese ships. She had uncontested authority over the pirates down the Moroccan coast and married the King of Morocco.[11]

The Barbary pirates particularly targeted Spanish ships and their raids were mostly routine: overpowering the ship, taking the crew to be sold into slavery and the goods to be traded. In one unusual incident a pirate ship went into a small Spanish port to rescue some Moorish slaves, 200 families in all. They rowed to the island of Formentera pursued by General Portundo and eight Spanish galleys. Unseen by the Spanish, the pirates landed the refugees on Formentera as the ships were too crowded to fight off the anticipated Spanish attack. The Spanish approached but held their fire as they believed the refugees to be still on the boats. They had been offered a handsome reward by the slave owner for the safe return of the slaves and so were fearful of drowning them in battle. The pirates then went on the offensive, rowing towards the Spanish fleet, killing General Portundo and overpowering the galleys. The refugees were re-embarked and the captured Spanish forced to row the pirates back to Algiers.[12]

Though operating through independent states, the Barbary pirates had links to the Ottoman Empire and recognised the Sultan Suleyman as the Imperial master. Barbarossa sent over lavish presents from his pirate raids including 'two hundred boys dressed in scarlet, each bearing a gold and silver bowl, two hundred more with rolls of fine cloth, and thirty to offer the Sultan thirty well stuffed purses'.[13] In a raid on Reggio in the Straits of Messina, Barbarossa took many prisoners including the Governor's 18-year-old daughter. He married her – even though, according to tradition, he was 90 years of age – and freed her parents as a wedding present. Barbarossa died in 1546 and his body was placed in a sepulchre that he had built. There were reports of the body rising from the grave. It was finally re-buried along with a black dog, which apparently stopped the levitations.[14]

Barbarossa's death did not signal the end of piracy. A few years later in Valencia, 'all the [regal] talk is of tournaments, jousting, balls and other noble pastimes, while the Moors waste no time and even dare to capture vessels within a league of the city, stealing as much as they can carry'.[15] Also at this time the wine trade between Provence and Corsica stopped, as 'the boats ... dared not put to sea, for fear of the twenty-three pirate ships prowling the coast'.[16] One source reports that in 1564 there were a hundred Barbary pirate ships operating and that it was 'raining Christians in Algiers'.[17]

In some respects, following Barbarossa's death the age of piracy was only just beginning. The pirates, however, needed the trading ships to be sailing in order to make a living, and there were periods such as 1560–65 when the Mediterranean was effectively closed to shipping because the pirate attacks had become so widespread. The pirates were forced to look for other battlefields. Algiers depended on them, the armies of slaves they provided, and the wealth from the stolen goods.

During the rise of the Barbary pirates, there was also a Christian presence. The historian Fernand Braudel is keen not to let the activities of the Barbary pirates who dominated the western Mediterranean block out the Christian pirates who operated mainly in the eastern Mediterranean, seizing cargoes of spices, silks, wood, rice, wheat and sugar.[18] Christian pirates attacked Christian ships as well as Turkish ones. In one pirate raid a Tuscan galley was seized by Venetians. The Duke of Florence proclaimed: 'Why should the Venetians have the right ... to prevent a Christian ship from putting out against the Infidel, if the said Christian vessel did not enter one of their ports? "Does not the sea belong to everybody?"'[19]

Markets in human beings captured by the pirates took place in Malta, Messina and Leghorn – all Christian ports – just as they did in Algiers. Sometimes Muslims captured by Christian pirates were swapped for Christian captives in Algiers.[20] Christian pirates sometimes also sold their stolen goods in Algiers. Braudel claims that the barbarity of the Barbary pirates has been exaggerated, and that they acted in good faith as often as did the defenders of Christianity. He believes that 'too much attention has been paid to the protests and arguments of the inhabitants of Christian shores and historians have sometimes been rather hasty in drawing conclusions ... Christian and Moslem piracy roughly balances out'.[21]

The pirates in the Mediterranean effectively asserted control over the sea by the power they wielded in attacking ships and crew and through their successful demands for protection money. 'All, from the most wretched to the most powerful, rich and poor alike, cities, lords and states, were caught up in a web of operations cast over the whole sea.'[22] Pirates snapped at the heels of legitimate traders. When European traders started venturing into the Atlantic on their way to the East Indies, the pirates were not far behind. By the early seventeenth century, the Algerians had learnt how to build square-rigged sailing ships and how to navigate them. This allowed the pirates to extend their raids into the Atlantic Ocean with attacks on the East India and Guinea traders.

The British were also threatened. One boat was captured by pirates in the Thames, and on the west coast of southern England the fear of piracy was so great that the light from the lighthouse off the Lizard Peninsula in Cornwall was extinguished because it could have been useful to pirates.[23] England, France, Spain, Holland and Sweden all attempted to pay pirates not to molest their vessels, without much success. Algiers became a phenomenon of worldwide international significance.[24] Every trading nation of the western world was affected.

Spain and Portugal were the first main European traders to try to reach the East Indies. The Portuguese sailed south along the west coast of Africa and then east reaching the East Indies. The Spanish ventured west and encountered the New World, a land later named the Americas after Amerigo Vespucci. These lands were claimed by the Spanish crown.[25] Spain and Portugal both attempted to protect the sea routes to these territories. Pope Alexander VI came to their assistance: in 1481 a Papal Bull granted Portugal 'a monopoly in the Ocean Sea towards the regions lying southward of the Canary Islands'.[26] This meant that Portugal owned the sea passage to the East Indies.[27] The attempt to close the seas would have aided the Portuguese in their fight against pirates. Any foreign ship in 'their' waters would now be fair game. Trading vessels going to the East Indies might be given the chance to pay the Portuguese for protection. If they failed to do so, they would be treated as pirates, their cargo taken, and the crew killed or sold as slaves.[28] In a Papal Bull of 1493 Spain was granted 'exclusive rights to trade and conquest across the Ocean Sea beyond a meridian 100 leagues west of the Azores'.[29] This is a stretch of the Atlantic Ocean totally 2,400 miles to the American coast. After some objection from the Portuguese, the Spanish sea was shifted 270 leagues further westward in the Treaty of Tordesillas. This received the Pope's endorsement.[30]

Spain and Portugal's trading rivals did not accept this assertion of ownership of the seas or the legitimacy of actions undertaken to enforce it. King Francis I of France, for one, strongly objected.[31] Also in retaliation, a Dutch East India Company boat captured a Portuguese trading vessel in 1604. The seized goods were sold in the Netherlands and profits distributed among the company shareholders. When Queen Elizabeth received complaints from the Spanish concerning Francis Drake's privateering activities, her response was that the Pope's divisions didn't bind her and that 'the use of the sea and air is common to all'.[32]

Paradoxically then, the attempt to close off the seas – partly initiated by concern over piracy – encouraged countries who had lost out in this closure to engage in raids on Portuguese and Spanish ships. However, a concerted attack on the North African pirates, the biggest threat, was not deemed prudent when these pirates could be counted on to weaken one's trade competitors.[33] Whether the raiders were called 'pirates' or 'privateers' seemed to depend on whose side you were on. Was there any clear way of distinguishing between pirates and privateers in the sixteenth century?

Privateering or hostile action undertaken by privately owned vessels in wartime thrived in the Mediterranean during this period. It was regarded as a legitimate form of war authorised either by a formal declaration, commission or instruction from a state or city.[34] Privateering involved the seizure of ships, goods and crew who were then sold into slavery – in other words, activities identical to those of piracy. As a form of war, 'Upstarts had replaced the tired giants [of the Islamic and Christian states] and international conflicts degenerated into a free-for-all' but the antiquity of the activity gives it a 'natural', 'human' quality.[35] Piracy or privateering in the sixteenth century 'consumed the passions that would in other times have gone into a crusade or *Jihad*; no one apart from madmen and saints was now interested in either of these.'[36]

Braudel claims that privateers were not people fighting for a cause. They were just making a living as sea-robbers, often seizing anything that came their way, sometimes even ships from their own side.[37] In such cases there is no question that privateers were pirates. There was also little restraint: 'anything was allowed – provided it succeeded'.[38] They used the markets in the Muslim and Christian pirate centres already mentioned. Registers of captives intending to pay ransom were kept along with lists of galley slaves for sale, or lists of people who could be exchanged.[39] Describing Algiers in the sixteenth century, Braudel writes: 'Privateering, the major industry, was the cohesive force of the city, creating a remarkable unanimity whether for the defence of the port or the exploitation of the sea, the hinterland or the masses of slaves.'[40] By the early seventeenth century Algiers was 'overflowing with wealth' and the city 'was now of a size to dominate the entire Mediterranean'.[41] So were even the Barbary 'pirates' really 'privateers'? The Dutch jurist Cornelius von Bynkershoek, writing in 1744, thought so. He said that the accepted definition of pirates was 'those, who, without the authorisation of any sovereign, commit depredations on the sea'.[42] The *Algerines*, *Tripolitan* and *Tunisians* are 'regularly organised societies who have a fixed territory and an established government'.[43] As they were acting with the authorisation of a sovereign state they could not be legitimately branded as pirates. Even if their deeds were worse than other seafarers, Bynkershoek thought that this was insufficient to brand them as pirates.

According to Thomson the distinction between pirates and privateers is meaningless, especially when basic questions about war and sovereignty are unanswerable.[44] By blurring the distinction between piracy and privateering, piracy could be seen in quite a positive light. Historically admired navigators were sometimes privateers. Sir Francis Drake sailed from England to attack Spanish colonies in America in the sixteenth century, plundering towns and seizing gold under orders from Queen Elizabeth I who called him 'my pirate'.[45] The Queen also had dealings with another pirate, the Irish woman Grace O'Malley, who had a fleet of 20 ships including several well-armed galleys. She led a band of over 200 men in what she described as 'maintenance by land and sea',[46] meaning theft and pirate raids. She was very successful. However, the English eventually captured her and her son. Grace was exchanged for hostages and appealed to Queen Elizabeth in person to release her son, promising to give up piracy. The Queen granted her request, but Grace broke her promise and resumed her 'maintenance by sea', asserting disingenuously that she was fighting the Queen's 'quarrel with the world'.[47]

In the seventeenth century, William Dampier travelled over vast distances, at first by hitching lifts on pirate ships, claiming he was 'more interested in what he was finding out about the world than the means by which he satisfied his curiosity'.[48] But Dampier wasn't just an innocent hitchhiker; he participated in pirate attacks against the Spanish. Also in 1703 he left England as captain of the *St George* and spent two years plundering ships in waters near South and Central America.[49] In the Gulf of Panama he abandoned his deteriorating ship, captured a Spanish one and sailed for Batavia.[50]

The regal defence of privateers stalled criticism. However there were strong voices against piracy. The sixteenth century jurist Alberico Gentili stated that pirates are common enemies, and can be attacked with impunity by all:

> Piracy is contrary to the law of nations and the league of human society. Therefore war should be made against pirates by all men, because in the violation of that law we are all injured ... No rights will be due to these men who have broken all human and divine laws and who, though joined with us by similarity of nature, have disgraced this union with abominable stains.[51]

The audacious attempt by Spain and Portugal to assert ownership of the Atlantic in the sixteenth century was in part due to the problems they were encountering with pirates. However, as noted above, their trading rivals, especially the Dutch, were unwilling to accept this domination of the ocean. The provocative seizure of a Portuguese trading vessel by the Dutch cried out for a legal defence. This was offered by Grotius, whose writings made him perhaps the most influential theorist concerning ownership of the seas right up to the present day. He provided a rigorous defence of the doctrine of the freedom of the seas which still influences international law.

GROTIUS' ARGUMENTS FOR FREEDOM OF THE SEAS

Grotius' treatise is called 'The Freedom of the Seas or the Right which belongs to the Dutch to take part in the East Indian trade'. It was published in Latin under the title *Mare Liberum* in 1608.[52] His main intent is to establish the right of navigation and the right to trade over the seas. To a lesser extent Grotius is concerned with the right to fish.

   Grotius is a skilful reasoner presenting many separate arguments which I have labelled below with headings for ease of reference later. He begins surprisingly with a circular argument stating that it is self-evident that 'Every nation is free to travel to every other nation, and to trade with it', and that therefore navigation is free to all persons. This is clearly inadequate as the premise was denied by the Portuguese. It was not self-evident to them. However Grotius' premise is backed up in his next argument:

The argument from God's intent

1. God wishes human friendships to be engendered by mutual needs and resources 'lest individuals deeming themselves entirely sufficient unto themselves should for that very reason be rendered unsociable'.

Therefore:

2. It is not God's will to have Nature supply every place with the necessities of life. 'He ordains that some nations excel in one art and others in another'.

Therefore:

3. By the decree of divine justice it was brought about that one people should supply the needs of another.
4. Those who deny this decree destroy this most praiseworthy bond of human fellowship. They 'do violence to Nature herself'.

Therefore:

> 5. God speaking through the voice of nature says that 'Every nation is free to travel to every other nation, and to trade with it.'[53]

Next, Grotius presents a delightful argument relating to the winds:

The argument of the winds

> The oceans are navigable in every direction by the 'regular and occasional winds which blow now from one quarter now from another'. Hence: 'Nature has given all peoples a right of access to all other peoples.'[54]

This argument is followed by two more short ones:

The argument from hospitality

> The law of hospitality demands that no 'state or ruler can debar foreigners from having access to their subjects and trading with them'. Hence the seas should be free to facilitate this access.[55]

The piracy argument

> Even if the Portuguese did own the trade routes, which Grotius is not granting, they would nevertheless be doing the Dutch an injury if they should forbid them access to the countries with which they want to trade. How much worse an injury are they inflicting on the Dutch given that they don't own the trade routes? In so far as the Portuguese, 'beset and infest ... trade routes, they are acting like pirates'.[56]

Next comes a key argument:

The argument concerning occupation

> 1. Sovereignty implies private ownership.
> 2. Private ownership arises from occupation.
> 3. The sea is limitless.

Therefore:

> 4. The sea cannot be occupied, so it cannot be privately owned.

Therefore:

> 5. There cannot be sovereignty over the sea.[57]

As noted below Grotius allows the possibility of ownership, even private ownership of some parts of the sea on the coastal fringe. However he intends his argument here to hold for nearly all the sea, and certainly all the open ocean. Attempting to counter a Portuguese response, Grotius says that simply sailing over the seas does not confer ownership. Otherwise previous sailors could claim ownership of the sea and exclude 'peoples of today', and there were previous sailors on the same routes. The ancient Romans regularly sailed from the Arabian coast to India and 'even so far as the golden Cheroneus which many people think was Japan'.[58] Grotius cites Pliny as saying that 'cohorts of archers were carried on the boats engaged in trade [with the East] as protection against pirates'.[59] So the Portuguese didn't discover the sea routes they claim to own and even if they did, then, as everyone knows, 'a ship sailing through a sea leaves no more legal rights than a track'.[60] Finally:

The common use argument

1. If something is so constituted by nature that although serving one person it still suffices for the common use then such a thing has been produced for the common use of man.
2. One person navigating or fishing the sea does not prevent others doing so.

Therefore:

3. The sea is for the common use of all men either for navigation or fishing.[61]

In a rhetorical aside Grotius says that preventing navigation 'amounts to cruelty' and 'If a man were to prevent others from fishing he would not escape the reproach of monstrous greed.'[62] He does leave an insightful loophole with fishing, however, stating that it may be possible to prohibit fishing 'for in a way it can be maintained that fish are exhaustible'.[63]

Other attempts to assert ownership of the sea are deflected by Grotius. He rejects the argument that the Portuguese were opening up the trade route to the East that had been interrupted for centuries which means they should be rewarded with ownership of the route. Grotius grants that if the efforts of the Portuguese benefited all humans then their position might be strong. However their efforts are exclusively for their own financial gain and other nations keen to be involved in trade with the East are kept out.

As I mentioned above, the Pope carved up the oceans between Portugal and Spain. Grotius regarded this an 'an act of empty ostentation'.[64] He believes that he has shown with the above argumentation that the sea cannot be the private property of any man and hence 'it could not have been given by the Pope nor accepted by the Portuguese'.[65] There are more arguments here which won't detain us, except for the final words on papal authority: 'the Pope has no authority to commit acts, repugnant to the law of nature' (see Grotius' argument from God's intent).[66] So he cannot allow private ownership of the sea.

Having established that the sea cannot become property, Grotius outlines how the sea cannot belong to the Portuguese by title of prescription or custom. Neither can hold as it is impossible to acquire by prescription or custom things that cannot become property. Also there is no one who is sovereign of the whole human race with competence to grant to any nation title to the sea by prescription as the sea is recognised as common to the use of mankind (see the common use argument).[67]

Custom is invalid when it is opposed to the law of nature (see the argument from God's intent). 'Custom is established by privilege. No one has the power to confer a privilege which is prejudicial to the rights of the human race.'[68] Grotius quotes Vasquez, a famous Spanish lawyer who argues against the position of his own country. Vasquez uses the rule, 'Whatsoever ye would that men should not do to you, do not ye even so to them', in order to reason that 'since navigation cannot harm anyone except the navigator himself, it is only just that no one either can or ought to be interdicted therefrom'.[69]

Concluding this whole complex of argumentation Grotius states: 'The Portuguese are in possession of no right whereby they may interdict to any nation whatsoever the navigation of the Ocean to the East Indies.' He then remarks that if a court judgment cannot be obtained to support Dutch interests then the Dutch should go to war as their cause is just.[70]

The Portuguese didn't withdraw their claims but nor did they offer any legal defence. Of course, if Grotius' arguments were valid they would work just as well against other nations asserting sovereignty over the sea, in particular Spain and Britain, and it was the British lawyers who rose to the challenge.

REPLIES TO GROTIUS DEFENDING CLOSURE OF THE SEAS

The two key responses to Grotius were by William Welwod from Scotland and John Selden from England. Welwod did not actually contest Grotius' claim that the high seas should be free. In fact he states that it is a 'ridiculous pretence' to argue for a liberty to sail the seas as this is uncontroversial. However, Portugal, Spain and John Selden do not agree. It can hardly be a pretence when Portugal and Spain try to keep other boats out of 'their' oceanic waters. Welwod believes that Grotius is really arguing about the right to fish the seas and discussing the 'right to sail' is a cover. This is quite an absurd claim when one looks at Grotius' text. Although he puts the right to fish alongside the right to navigate, he acknowledges that there could be grounds for limiting fishing, as mentioned above. Welwod's case for private ownership of fishing rights is not strong. It draws on the belief laid down in Genesis that God said 'Subdue the earth and rule over the fish'.[71] Welwod is concerned about the foreign fishers of white fish off the east coast of Scotland, how they scatter the fish away from the shores so that 'no fish can be found worthy of any pains and travails, to the impoverishing of ... our home fishers and to the great damage of all the nation'.[72] Grotius replies that when God says rule the fish, God is pronouncing to the 'whole human race' so it does not justify any claim to private fishing rights. 'There is no question there of a right which is competent to men against other men, but of one to all men against the lower creatures.'[73]

The most powerful response to Grotius was from John Selden in his book *Mare Clausum* (Closed Seas), first published in 1635. Selden tackles Grotius' arguments more directly. Against the argument from God's intent, Selden says that nations have become more self-sufficient.[74] However this is only part of the argument. Selden does not engage with Grotius' claim about sociability. He looks at the argument from occupation and challenges the premise about the limitless nature of the sea on the basis that the sea is bounded by shores. He admits that it is a little more difficult to find limits or bounds in 'the main Sea' but he says that rocks and islands could be used. Also the compass, latitude and longitude can be drawn upon to create boundaries.[75] (It took over 100 years, however, to get anything like an accurate measure of longitude.) Selden picks up on Grotius' point that if any small part of the sea can be enclosed and occupied then it can become private property. He goes on to ask, what difference does it make if we talk about a small part of the sea or a larger part? He argues that if the larger part could be occupied then Grotius would have to admit it could become private property.[76] Grotius does not offer any replies to Selden. He could, however, give up the premise that the sea is limitless and still claim that it cannot be occupied. Being able to theoretically measure areas of the sea is not the same as being able to occupy them. Likewise, Selden could have tightened his case by disputing the need to occupy in order to establish ownership.

Selden's two long books on the defence of closed seas consist largely of records of customs, and accounts of how various nations have claimed sea territory throughout history, and various rulers taken for themselves the title of 'lord of the seas'. He states that such ownership was initially based in Divine decree when God invested Noah 'in the Dominion of the whole Earth (of which Globe the Seas themselves are a part)'.[77] Since then the laws of diverse nations have taken over to justify possession. Selden is not just concerned with coastal seas. He addresses Portugal's claim to the Atlantic. He states that it is forbidden for non-Portuguese ships to enter there and all Portuguese commanders can call foreign shipping in the Atlantic Ocean into account. 'So that we see the Nation of Portugal also made no question, but that Dominion might be justly acquired over the Ocean itself.'[78]

The Portuguese may make no question, but Grotius has and other nations did too, nations that also wished to trade these routes. Grotius' argument about custom and prescription being an inadequate basis for the justification of ownership of the sea is simply not addressed by Selden. Selden's position, while purporting to be a general defence of ownership of the sea by various nations, is more accurately seen as an argument for British ownership of the British Sea. In the dedication at the beginning of *Mare Clausum* he says: 'What true *English* heart will not swell, when it shall be made clear and evident (as in this Book) that the Sovereignty of the Seas flowing about this Island, hath, in all times ... down to the present Age, been held and acknowledged by all the world.'[79] In a direct verbal challenge to Grotius, Selden says in his Preface that 'the sea, by the law of nature or nations, is not common to all men, but capable of private dominion or property as well as land; ... the King of England is Lord of the Sea flowing about, as an inseparable and perpetual appendant of the British Empire'.[80] The geographical extent of the British Seas was vague, especially to the west. They expanded and contracted with the ability of Britain to defend them. A map of the British Seas claimed in 1635 is reproduced in Figure 1.2.

Like Protagoras, Grotius could argue for two sides of a case. A few years after *Mare Liberum* he offered a defence of the closure of the sea to suit Dutch interests. The Dutch had gained territory in the East Indies, in particular the Moluccas or Spice Islands. Grotius

Figure 1.2   The British Seas in 1635. Source: J. Selden, *Mare Clausum: Of the Dominion, Or, Ownership of the Sea*, New Jersey: The Lawbook Exchange Ltd, 2004 (1652), unnumbered.

argued for the closure of the seas around these regions, apparently with 'uncommon ability'.[81] Boxer claims that the oppressive commercial monopoly of the Dutch East India Company led to a rise in piracy carried out by people from the Spice Islands.[82] It is the arguments of *Mare Liberum*, however, which have had a more lasting influence.

Grotius made use of ideas already expressed by two Spanish writers, which is interesting since Spain supported closure of the seas. One was the above mentioned Vasquez, who put forward the argument from occupation before Grotius.[83] The other a Spanish monk, Francis Alphonso de Castro, who objected to the Venetians closure of the Adriatic Sea in words very similar to the argument from God's intent.[84] These two authors, however, could not question the Papal Bulls and nor did they offer such a strong appeal to a sense of justice as Grotius provided. In European waters during the seventeenth century the state practice favoured the freedom of the seas, and by the eighteenth century this was accepted more generally. Throughout the following chapters I will be noting the influence of Grotius' principles on contemporary issues concerning the sea, and in Chapter 8 I will show specifically how this heritage is still influential in western legal judgments. As James Brown Scott says when introducing the 1916 translation of *Mare Liberum*, Selden's treatise is 'heavy and water-logged' and has 'gone under', whereas Grotius' doctrine 'rides the waves'.[85]

LIMITS TO FREEDOM OF THE SEAS

When the claim is put that the sea should be free, is it the case that every part of the sea is included? Grotius goes backwards and forwards on this. First he says that all the sea, the sands of the sea and the shore should be free since they are susceptible to universal use.[86] Then he argues that if any part of the sea or shore can be occupied it may become the property of the occupier 'so far as such occupation does not affect its common use'.[87] He allows structures to be built on the shore if they do not inconvenience others. However 'the sea seems by nature to resist ownership' and may sweep in and recover possession of the shore.[88] Fencing off an inlet for private ownership is also allowed by Grotius. He relates Pliny's story of Lucullus who 'brought the water of the sea to his villa by cutting a tunnel through a mountain near Naples thereby creating private fishing ponds'.[89] Grotius does not consider such a move as hindering the common use of the sea.

For Grotius, harbours and inner seas cannot be privately owned but can be owned by the nation.[90] Gulfs and straits and 'all the expanse of sea which is visible from the shore' are also not included in his plea for the freedom of the sea.[91] The plea relates to

> The OUTER SEA, the OCEAN, that expanse of water which antiquity describes as the immense, the infinite, bounded only by the heavens, parent of all things; the ocean which the ancients believed was perpetually supplied with water not only by fountains, rivers, and seas, but by the clouds, and by the very stars of heaven themselves; the ocean which, although surrounding this earth, the home of the human race, with the ebb and flow of its tides, can be neither seized nor enclosed; nay, which rather possesses the earth than is by it possessed.[92]

In the uncertainty as to what exactly counts as this 'OUTER SEA' or 'OCEAN' Grotius has paved the way for limitations on the freedom of the seas.

An English proposal was that the high seas began beyond where a cannon shot could be fired.[93] The rationale was that this range of sea territory could be protected by cannons on the land. Other coastal nations agreed to this rule for fixing territorial waters, setting the limits to national sovereignty. In 1793 the Americans claimed a three mile territorial sea. This was taken up by the British. However a major inroad into the 'OUTER SEA' did not occur until 1945 when President Truman laid claim on behalf of the US to the continental shelf and coastal fisheries off the US coast. A new carving up of the oceans began making serious inroads into Grotius' outer seas. Fishery limits remained vague until 1976 when the exclusive US fisheries zone was set at 200 miles. Other nations followed, leading to conflict. The United Nations General Assembly became involved and initiated a series of conferences and conventions on the Law of the Sea. In the latest Convention (LOSC),[94] which entered into force in 1994, there is agreement on a 12 mile territorial sea and a 200 mile Exclusive Economic Zone (EEZ).

Every coastal state can lay claim to the waters up to 12 nautical miles out. The coastal state has sovereignty in their territorial sea with, however, one major limitation: the right of foreign vessels to innocent passage. The EEZ may extend out another 188 nautical miles from the territorial sea. The coastal state does not have sovereignty over this zone but it does have particular sovereign rights and jurisdiction. It has sovereign rights to the living resources in the water column, and the non-living resources in the water column, seabed and subsoil. It also has sovereign rights to other activities for the economic exploitation and exploration of the EEZ. The notion of sovereign rights here is somewhat confusing, as is the notion of exclusivity, because there are obligations to share fisheries, for instance, where surplus stocks exist. The coastal state also has jurisdiction over research and environmental management in the EEZ. This includes obligations to protect and preserve the marine environment within this area.

While some coastal states are agitating to push the EEZ even further out (and archipelagos create a different complexity) in general the area beyond the 200 nautical mile zone is the high seas. This area of the sea is still free. Coastal states can make claims on the seabed and subsoil of submarine areas of continental shelves, for example to exploit oil and gas, from 200 to 350 nautical miles, but this does not mean that the coastal states have a claim on the ocean above the continental shelf. Marine areas beyond the 12 mile territorial sea have become known as 'international waters'. This label captures the lack of sovereignty in these waters but it is misleading given the coastal states' rights, obligations and jurisdiction.

GROTIUS' PRINCIPLES IN THE CURRENT LAW OF THE SEA

The seas which are free are now much smaller than Grotius envisaged. However two key notions are still important: the right to navigate and the right to fish. For Grotius 'The same principle which applies to navigation applies also to fishing, namely that it remain free and open to all.'[95] With the new domains in the sea these rights become qualified. On the high seas they generally still hold. There are some uneven attempts at restrictions by regional agreements to be discussed in Chapter 5.

Legally any state has the right to fish in its territorial sea and EEZ and the EEZ of foreign states if there are surplus stocks. However it is the coastal state that determines whether there are surplus stocks. Coastal states usually attempt to keep foreign fishers out. In practice this may be difficult especially in remote areas, for instance in the EEZs off Heard Island and McDonald Island in the Southern Ocean, or where governments are unable to provide policing such as in Somalia.

Any state has the freedom of navigation through the EEZ of another state subject to the laws of the coastal state. Any state has the right to innocent passage in any territorial sea. 'Passage is innocent so long as it is not prejudicial to the peace, good order or security of the coastal state.'[96] This provision can be used to further close off the sea if the coastal state declares maritime security issues are at stake, perhaps without warrant.

Concerns over environmental security have been used to limit navigational rights though these acts could amount to denying innocent passage. Some states by their actions are challenging the extent of the right of innocent passage in the Law of the Sea. Indonesia, Malaysia and Singapore refused permission for the *Akatsuki Maru* that was carrying plutonium to pass through the Straits of Malacca and Singapore in 1992. Concern over environmental security was used as a basis to limit navigational rights.[97]

In 2004 fuelled by a fear of terrorism and the need to protect shipping, ports and oil rigs from attack, Australia set up the Australian Maritime Information Zone which extends to 1,000 nautical miles from the coast.[98] Ships within this zone are required to provide details of their journey and what they are carrying. When ships enter the EEZ they need to give more detail of the cargo, ports visited, shipowners, registration and destination. The Border Protection Command introduced in 2005 is able to 'independently order the interception of ships within the information zone'.[99] Foreign shipping is restricted more in the Australian EEZ than is laid down in the Law of the Sea. These new initiatives amount to an attempt to restrict navigation for a further 800 nautical miles. The Australian EEZ is already over 8 million square kilometres, making it one of the largest EEZs in the world.[100] Indonesia strongly objected as the new Australian zone would take in seas that 'come totally under Indonesian jurisdiction' and it violates international law. Australia's response was that they wanted to protect their own interests.[101] Negotiations continue. New Zealand ministers also expressed alarm at the possibility that Australian ships might try to intercept foreign ships in New Zealand waters.[102]

As more and more of the sea is closed off it is timely to ask whether any of Grotius' arguments for the freedom of the seas still have force. The argument from God's intent is not likely to be taken seriously today. The common use argument has been overridden by new considerations. The right of navigation may be challenged, for instance when vessels are carrying dangerous cargo. Issues of environmental security would not have arisen 400 years ago and yet they may form a reasonable ground for denying passage.

Piracy is on the rise again now, especially in the Gulf of Aden and the Somali sea. Up until 2008 attacks took place mainly in coastal waters off states unwilling or unable to counter piracy. Hijacked vessels 'disappeared' on the high seas and there was great reluctance to pursue them given that no central authority existed to give backing to pursuit and prosecution. From 2008 Somali pirates have made many attacks on the high seas. In the early stages they were able to make use of the Somali territorial sea to evade capture by foreign vessels. Now these vessels have been given the right by the UN to enter Somali waters. There are difficulties in pursuing pirates in any waters and some of these difficulties arise because of the right of navigation in particular for the mother ships servicing pirate activities. This is only one part of the answer to the question, how did the Somali pirates get into a position of ruling the waves? This question along with the ramifications for ocean governance will be taken up again in Chapters 4 and 9.

Welwod's concern about the exploitation of fish stocks foreshadowed the current critical state of fisheries. By the early 1990s most stocks of commercially valued fish were running low, according to the Food and Agriculture Organisation of the United Nations, and wars have broken out over fish populations (these issues are discussed in Chapter 5 below). In 2009 the European Commission proposed a moratorium on catching white fish off the west coast of Britain to help fish populations recover after years of overfishing, the same species that Welwod wrote about in the seventeenth century.[103]

For Grotius, keeping the seas free was a question of justice. In the present day, we have a much greater capacity to bring about environmental destruction. Given the right of all to fish the high seas and very limited pollution controls, this seems to be exactly what is happening. Sylvia Earle is a scientist who has published prolifically, framing many of the discussions and policies concerned with how to protect the oceans in the last decade. She argues that not only human well-being depends on ocean health but our very survival: 'if the sea is sick, we'll feel it. If it dies, we die. Our future and the state of the oceans are one.'[104] Grotius was concerned with what was just for the existing people. Justice however also requires us to keep in mind future generations. So the freedom of the seas may need to be curtailed to foster environmental protection to save something of the oceans' health and wealth for future generations.

Grotius' argument concerning occupation still has force. However the distinction between sovereignty, sovereign rights and jurisdiction means that use of the sea can be restricted while the sea remains unowned. The twentieth century saw the establishment of an international Law of the Sea. The legislation grants to coastal states sovereignty in the territorial seas but not beyond, which is probably within the bounds of what Grotius accepted. However the Law of the Sea has endorsed broader and broader legal control by coastal states beyond the territorial seas, whether it be in the form of sovereign rights or jurisdiction. Grotius' premise that the sea is limitless is false, and newer measuring instruments mean that sea borders, though politically contentious, are geographically safe. A 200 nautical mile incursion into the high seas is contrary to Grotius' intent even if it doesn't involve sovereignty.

One of the factors motivating Grotius' position was that there was no justice in dividing up the Atlantic Ocean between Spain and Portugal and attempting to exclude other nations. In the legal extensions that are now occurring a similar injustice might result. The coastal states, especially those with long sea borders, stand to gain greatly from the living and non-living resources in the EEZs, for example, fish off northern Australia, or oil off the west and north coasts of Norway. Land-locked states may well feel disadvantaged. It is no longer necessary to have a seaport to exploit sea resources but land-locked states do not have the same entitlements as coastal states. Also, some states are pushing for an even broader legal reach as in the Australian example above. These encroachments into the sea space beyond EEZs are not assertions of ownership but they have the potential to generate the same sort of discontent that assertions of ownership of the sea had in the past, cries that this is unjust and that resources are unfairly divided.

Taken together with the considerations about the harm being done to the oceans with a free seas regime, the best way forward may be to advocate sovereignty of the seas beyond the 12 mile limit to a central organisation allied to the UN, perhaps funded by fishing licences, licences to exploit the minerals in the seabed and licences to navigate, to adequately cover the fight against piracy and the environmental management of the oceans.

In the twenty-first century consideration of who owns what in the sea needs to be taken in the context of climate change and possible submersion of lands.

CLIMATE CHANGE, RISING SEA LEVELS AND THE DISPLACEMENT OF ISLAND COMMUNITIES

Climate change will impact on the sea territory of inundated islands and mainlands. If the ocean moves in over mainlands then borders of the territorial sea and EEZs may need to be re-drawn. If water moves across low lying islands that then have to be abandoned, the sea territory as well as the land territory of those communities is lost. Sea refugees will face enormous social, cultural and economic upheaval. They will need to depend on the assistance of other states to provide land or to sell land to them.

Twenty years ago the Maldives, an island nation in the Indian Ocean, was considered to be at risk of sea-level rise in the long term. The Maldives consist of 1,192 islands often with reclaimed areas, or atolls. The capital, Malé, is generally about two metres above sea level. Tulhadoo is only 35 centimetres above high water in the reclaimed areas, otherwise double that height. Addu Atoll is approximately 70 centimetres above high water. Thaa Atol is even lower. Some other islands such as Fua Malaku are higher. However, relocating people from the low-lying islands to the high ones may be unsuccessful as storms carrying salt water over even these islands disrupt agriculture and fresh water supplies. Nevertheless a 1989 report states that 'national resettlement efforts should focus on islands with substantial elevation',[105] namely, approximately two metres. Given that most climate models predict about a one metre rise in sea level by 2100,[106] it is unlikely that such a policy would be sufficient. The situation could be a great deal worse. According to the UN 2007–8 Development Program Report, 'Accelerated disintegration of the West Antarctic ice sheet could multiply sea-level rises by a factor of five.'[107] In November 2008, the Maldivian President Mohamed Anni Nasheed announced in his inauguration speech that he would set up a 'sovereign fund' for the relocation of the approximately 300,000 residents, possibly to India, Sri Lanka or Australia, if sea-level rise made such a move necessary.[108]

Some other countries are preparing to abandon their land and sea territory with the expected sea-level rise. The Government of Tuvalu, a low-lying Pacific island nation has made an arrangement with New Zealand whereby half of the population of 10,000 will move to New Zealand and work in agriculture if Tuvalu is swamped.[109] The Seychelles, another Indian Ocean island nation, is also concerned about the likely loss of 60 per cent of their islands due to sea-level rise.[110] Kiribati in the Pacific Ocean has 32 atolls that face submersion and one island peaking at 6.5 metres above sea level.

The ecological impacts of climate change may compound the problems of sea-level rise making low-lying islands in the tropical regions uninhabitable. Warmer oceans can cause bleaching of corals.

Storm surges that are on the increase further degrade the corals weakened by bleaching. Ocean acidification caused by rising CO_2 levels can also destroy corals. The fish that depend on the corals die off, thus threatening food security. This is already evident in the Maldives where coral reef decline has led to the die off of baitfish that are used in tuna fishing. Tuna fishing is now affected.[111]

These are just some of the impacts of climate change. The island communities likely to suffer the most are generally not part of the industrialised world that produced this crisis. The responsibility for relocation should not fall solely on the sea refugees. The international community through the UN should take the initiative in finding a way out if relocation is necessary. Even if the plight of low-lying island communities was the only consequence of climate change, it should provoke a re-think of the fossil fuel industries that are now thought to be largely responsible for the problems. Huge reserves lie under the sea. There are compelling reasons to leave them there.

2
Underwater Non-living Resources

With the Arctic the stakes are global.
M. Sommerkom and N. Hamilton, *Arctic Climate Impact Science*[1]

The earth's resources that lie under the sea are becoming more accessible with technically sophisticated shipping and equipment. Global warming is unfreezing parts of the polar seas also opening up possibilities for undersea mining. Questions about the economic desirability of extracting these resources are set against the desirability of retaining wilderness areas and the need to protect the oceans from the polluting effects of the mining industries. Profound results for international relations and even for life on the planet may hinge on whether to proceed with the extraction of fossil fuels in the Arctic and Southern Ocean seabeds.

WHO HAS A CLAIM?

Within the 200 nautical mile EEZ coastal states have sovereign rights over undersea resources such as oil, gas and minerals. Some states choose not to exercise these rights. For instance, since 1981 there has been a ban on drilling offshore around the US (except in the western and central Gulf of Mexico, offshore Texas, Louisiana, Mississippi, Alabama and Alaska).

Exploration beyond the EEZ has been difficult, but exciting discoveries in the 1960s revealed great wealth. Arvid Pardo, the Maltese Ambassador to the UN, informed the General Assembly about the rich mineral deposits lying on the floor of the central Pacific and Indian Oceans. These are called 'manganese nodules' and they contain valuable metals such as nickel, manganese, copper and cobalt. This announcement led to discussions concerning who should have ownership over this wealth. The US played a leading role, arguing that these minerals should be exploited for the benefit of all rather than just the developed states with the technical resources to find and exploit the minerals, given that the resources were lying in the seabed of the high seas. An International Seabed Authority

29

was to be set up to administer the sharing of resources. This idea is encapsulated in the Law of the Sea (Part XI).[2] However, as the US moved through different political regimes the principle of sharing exploitable resources in the high seas became less and less palatable. Since President Reagan's administration in 1981 there has been a preference for US mining to be for US benefit only. It was largely on these grounds that the US declined to ratify the Law of the Sea.[3] Other states tried to be accommodating by drastically weakening the need to share mining profits and cutting back the powers of the International Seabed Authority. This led to the Agreement that came into force in 1996. However these concessions were still not enough and as of 2009 neither the Law of the Sea nor the Agreement have been ratified by the US.

The US has until now been instrumental in blocking any attempt to vest ownership of undersea non-living resources in common heritage as opposed to private companies that exploit the sea floor of the high seas under the freedom of the seas doctrine. The International Seabed Authority (ISA) became operational in 2001. It is an autonomous body based in Jamaica attempting to administer a mining regime for the high seas areas that leads to an equitable use of resources. It is particularly focused on polymetallic sulphides, formed around hot springs in active volcanic areas, and cobalt-rich crusts, fused to the underlying rock around ridges and seamounts in all the world's oceans.[4] In 2009 eight countries had contracts with the ISA. When mining becomes profitable the contractors will pay royalties to the ISA who will then distribute them fairly, taking into account the interests and needs of developing countries.

The Authority is limited in its powers and scope. It has been mainly concerned with minerals but aims to develop a code that will cover all seabed resources. Oil, gas and precious metals are of particular current interest. Its powers can be increased if there is the political will to do so. There are strong indications that the US is now considering ratifying the Law of the Sea. David Sandalow is a Senior Fellow at the Brookings Institution, a non-profit public policy organisation based in Washington DC. He claims that there is support for this ratification from the Joint Chiefs of Staff, the navy, the oil and gas industry, the fishing industry and major environmental groups.[5] In 2007 President George W. Bush urged the Senate to ratify, stressing that it would give the US sovereign rights over extensive marine areas.[6] If ratification of the Law occurs the Agreement would have to be ratified as well because it constitutes a legally binding modification of Part XI of the Law of the Sea.

These moves by the US could amount to a financial boost to the ISA through mining royalties. David Sandalow also acts as a foreign policy advisor to the US Government. He notes that the oil and gas industry does not oppose the modest revenue sharing provision in the Agreement. However, in a more sinister comment he says that if the US ratifies the Law of the Sea, then

> the United States would be a permanent member of the Council [governing the ISA] and have the ability, under relevant voting rules, to block most substantive decisions of the Authority, including any decisions with financial or budgetary implications and any decisions to adopt rules, regulations, or procedures relating to the deep sea mining regime.[7]

This is hardly the spirit of international co-operation that the ISA hoped to inspire. However the new US administration led by President Barack Obama may bring a change in perspective.

There is one oceanic domain receiving special treatment under the Law of the Sea and it is an area of escalating conflict. This is the area between 200 and 350 nautical miles out from the coast where a state's continental shelf extends that far.[8] If a coastal state can prove the extension then it has sovereign rights to explore and exploit the natural resources on the seabed and in the subsoil.[9] The notion of sovereign rights is not the same as sovereignty. So while strictly speaking the coastal states do not own the seabed of the continental shelf, the practical reality is indistinguishable from ownership. In addition to the right to exploit, the coastal state has the right to exclude others exploiting the area without their express consent. A small but potentially important consequence of the discussions concerning sharing the wealth from resource extraction in the high seas exists in the Law covering this 200 to 350 nautical mile zone. Article 82 of the Law of the Sea requires the sharing of the benefits of exploitation of non-living resources with the least developed and land-locked states that are signatories to the Law of the Sea. The sharing arrangement begins six years after production starts and in that year 1 per cent is paid to the ISA for distribution. This increases to 7 per cent over the next six years and then stays at that level.

While the intentions of this Article are good, there are some problematic features. Working out how far a continental shelf extends can be geologically challenging. A continental shelf is a natural prolongation of the landmass of the coastal state. It usually extends to the top of the continental slope at the outer edge of the continental margin. It does not include the deep ocean floor.[10] However this simple image can be complicated by faulting, slippage, and by sedimentation. Even with good underwater mapping equipment there is still the possibility of disputes about how far a continental shelf extends from the coast of a state. In addition, if it is fair to share the benefits of exploitation at 201 nautical miles from the coast it is also fair to share such benefits if the exploitation takes place 199 nautical miles out. The 200 nautical mile line is arbitrary and is made even more so by the legal fiction of 'baselines'. These are the lines drawn around coasts from which measurements are taken. They are straight, joining headlands for instance. This is a very rough measure, especially with ragged coastlines and offshore islands. If baselines are used, exploitation may be within 200 nautical miles, but if 'from the coast' is a more fine-grained determination then the exploitation might be beyond that line.

Developments in offshore exploitation could happen quickly. The search for these resources is greatly enhanced by the new equipment also used to look for underwater cultural heritage. The melting of the polar ice caps make the task easier in the Southern Ocean around Antarctica and in the Arctic Ocean, even though extreme weather conditions still pose difficulties. It is in these two areas that we are likely to see the escalation of international conflict. The race to exploit undersea resources is on.

ANTARCTICA AND THE SOUTHERN OCEAN

The Antarctic Treaty signed in 1959 – originally by 12 nations and now by 46 – aims at the common good. The guiding principle is environmental protection. The Treaty prohibits oil, gas and mineral exploitation in Antarctica until 2048.[11] The UN told states that if they want to make claims for jurisdiction over a continental shelf extending from their coasts then they needed to do so by May 2009. This applied not only to states with territorial claims in the Antarctic, though it has been taken up by those states. Britain has claimed more than 1 million square kilometres. This is challenged by Argentina, since British and Argentinean Antarctic territories overlap.[12] Australia, Norway and New Zealand have also made claims on the continental shelf areas in Antarctic waters. France has put in a claim for the seabed out from the Kerguelen Islands near Antarctica and a 400,000 square kilometre expanse of ice known as Adélie Land.[13] Staking a claim now could open up the possibility of underwater resource exploitation after 2048.

One of the problems with these moves is that they weaken the Antarctic Treaty. They are inconsistent with the Treaty's statement that 'No new claim or enlargement of an existing claim, to territorial sovereignty in Antarctica shall be asserted while the present Treaty is in Force.'[14] Some states that are now going against this agreement had re-asserted its importance until very recently. The Australian parliament announced in 1987 that there should be a long-term continuation of the 'freeze' on territorial claims in the Antarctic.[15] If the UN tolerates such an inconsistency then it is possible that the prohibition on mining might be lifted too. This may appear overly cynical, nevertheless a recent report in the *Guardian* revealed that 'The Foreign Office ... has told the *Guardian* that data is being gathered and processed for a submission to the UN which could extend British oil, gas and mineral exploitation rights up to 350 miles offshore into the Southern Ocean.'[16]

The preamble to the Treaty speaks of the promotion of international harmony. Allowing nations to individually prepare to exploit undersea resources will work against that aim. The Treaty also strives to preserve and conserve living resources in Antarctica in Article IX. This is likely to be inconsistent with seabed mining, as I will illustrate below.

THE ARCTIC OCEAN

The competition to claim territory in the Arctic Ocean is much more intense. This Ocean does not enjoy the same protection that is currently afforded to the Southern Ocean. Also the manner of the ice-cap melt in the Arctic opens up possibilities for mining more easily than around Antarctica. The resources available in the Arctic Ocean seabed and subsoil are thought to be huge. When Russia used a submarine to plant a titanium flag 4 kilometres below the ice of the North Pole this was a piece of theatre. However the act was also used to assert ownership of the ocean floor based on the claim that the undersea Lomonosov mountain chain links Siberia with the Arctic: 'the Lomonosov Ridge is the same nature as the continental shelf' according to Valery Kaminsky, the Director of the Russian Maritime Geological Research Institute.[17] This view is contested. Denmark and Canada claim that the Lomonosov Ridge is an extension of the North American continental shelf into the Arctic.[18] The US is working to support a claim on the continental shelf north from Alaska around the Chukchi Cap.[19] Norway also has interests in Arctic Ocean exploration and there is a natural gas plant already running in the sea off Hammerfest, Norway's most northerly town.

Denmark's claim is made through Greenland, which it administers. For its part, Greenland has seen the possibility of independence in the pursuit of oil. Denmark could benefit from undersea mining in a profit-sharing arrangement between the two countries. Alternatively Denmark stands to gain from Greenland's independence as Greenland is currently receiving subsidies from Denmark that would then cease. The estimates of the potential oil deposits that could be claimed are very large, with one area alone off northeast Greenland estimated to have oil and gas resources of 31 billion barrels of oil, an amount approximating the total of the world's consumption for a year.[20]

The fact that the Law of the Sea offers the possibility of making a claim for territory 350 nautical miles offshore is feeding the build up of international tension in the Arctic. In 2008 Russia began reinforcing its military presence in the Arctic Ocean. The Russian General Shamanov said that 'Russia had the capability ... to defend its claim to roughly half of the Arctic Ocean – including the North Pole.'[21] Canada too has intensified its military presence in the area and the Canadian Prime Minister, Stephen Harper, has unveiled plans for an army training centre for cold-weather fighting to be built in Resolute Bay, Nunavut, approximately 600 kilometres from the North Pole.[22] Norway expanded its high north military facilities in 2007 and is increasing military exercises in the north. In addition, Norway has entered into dialogue with China and Japan over the Arctic.

The EU became involved in 2008, led by a speech from the Fisheries and Maritime Affairs Commissioner Joe Borg.[23] While expressing a concern for the fragility of the Arctic, he also mentions the enormous hydrocarbon resources that are now more accessible with the ice retreat. He supports the sustainable use of these resources. In a subsequent meeting of the European Commission it was stated that 'the EU should ... work with Norway and Russia to exploit [the Arctic's] untapped hydrocarbon reserves in an environmentally sustainable way'.[24] Denmark, through Greenland, is the only EU country with a claim on the Arctic seabed. Norway is not a member of the EU. So exactly what is envisioned by way of EU involvement is hazy. It could be read as the EU seizing the chance to take regional responsibility, or as a move to try to keep Canada and the US out of Arctic affairs and to support natural resource exploitation for the benefit of Europe and Russia. The EU suggestion has not been taken up with enthusiasm by Canada who, following the announcement, re-enforced its jurisdictional claims on the Arctic. Talking about the Canadian Arctic, Prime Minister Harper said 'we are ... going to put the full resources of the Government of Canada behind enforcing that jurisdiction'.[25] Late in 2008 the EU announced plans to build *Aurora Borealis*, a new type of icebreaker capable of drilling under the sea floor up to 1,000 metres at a water depth of 5,000 metres in icy conditions.[26] Although this ship is designed as a research vessel, the knowledge gained about undersea geology would be invaluable to mining operations. So again, it should be asked: is the EU simply interested in science or is there an economic agenda?

Thus, instead of promoting international co-operation, the Law of the Sea encourages international competition, and competition based on a geological determination that may always be difficult to establish with any certainty. It gives the Arctic Rim states a vast quantity of the oceans' resources, locking out the rest of humanity. Also, allowing these states to make claims to a distance of 350 nautical miles acts as an incentive to exploitation of undersea resources that has the grave potential for environmental pollution in environments that are relatively pristine, and that are undervalued by the majority who live in a temperate world.

ECOLOGICAL THREATS FROM OIL AND GAS ACTIVITIES IN THE ARCTIC

The best basis for looking at these threats is the massive scientific report published by the Arctic Council in 2008,[27] and the studies done on the *Exxon Valdez* oil spill. The Arctic Council member states are Canada, Denmark (including Greenland and the Faroe Islands), Finland, Iceland, Norway, Russia, Sweden and the US. The Report is entitled an 'Oil and Gas Assessment' and it reviews the available scientific data in several hundred pages. The focus is on the Arctic but many of the discussions are relevant to the Southern Ocean as well. According to the report there are dangers in all oil and gas activities in the Arctic, in exploration as well as extraction, in transportation and in its use, for example, in electricity generation.

Exploration requires the construction of infrastructure that can degrade the surrounding environment. There can be oil discharges from the top of the borehole and from non-oil-based drilling fluids put into the water from offshore operations. There are negative impacts from decommissioning. Large-scale marine operations are needed to remove installations. These operations produce greenhouse gas emissions.[28]

During extraction 'produced water' is released into the surface of the ocean. This water comes from the subterranean oil-bearing rock formations. It may contain metals, salts and radionuclides, dissolved hydrocarbons and dispersed oil that can have toxic effects on marine ecosystems.[29] Workers are often exposed to drilling muds at offshore oil and gas operations. These contain similar mixtures as produced water that can be toxic to marine ecosystems and humans.[30] Well gas contains methane, butane, carbon dioxide and hydrogen sulphide, the gases which produce the greenhouse effect. Hydrogen sulphide is of particular concern for human health.

Leaks and spills of crude oil can occur during offshore operations. A blowout is where the well flows uncontrollably. It is sometimes possible to drill a relief well, but this is not so easy under the ice. In 1977 there was a blowout from the *Bravo* drilling platform in the Norwegian North Sea. In this incident $12,700m^3$ of oil was released into the sea.[31] The grounding of the *M/V Selendang Ayu* in Alaskan waters in 2004 led to a fuel spill where a $5km^2$ area of water was covered in emulsified oil.[32] This is a mixture of oil and water also called 'chocolate mousse'. The cold temperatures make oil susceptible to the formation of mousse. The volume of the oil slick increases with emulsification by a factor of two to five.[33] This process also slows down weathering of an oil slick, especially by evaporation or natural dispersion. Such weathering helps to get the oil out of the ecosystem.

The *Exxon Valdez* spill in 1989 – on the edge of the Arctic in Prince William Sound, Alaska – had profound, long-term effects on the marine ecology. The crude oil in this spill quickly turned into chocolate mousse. A study into the effects of the *Exxon Valdez* oil spill, reported in 2003, pointed to the 'unexpected persistence of toxic subsurface oil and chronic exposures, even at sub-lethal levels, [that] have continued to affect wildlife. ... Population reductions and cascades of indirect effects postponed recovery.'[34] The report exposes the weakness of approaches which look at species in isolation from each other. It documents indirect interactions traditionally overlooked, for example, the rockweed *Fucus gardneri* was destroyed along rocky shorelines in the spill. This provided protective cover for invertebrates. A green algae took over the space along with an opportunistic barnacle *Chthamalus dalli*. It was over a decade before *Fucus* could return and invertebrates re-colonise.

The Arctic Council report acknowledges that in 2008 problems from the spill continue.[35]

Russia is more involved in oil and gas operations than other Arctic countries and many strandings and collisions of tankers have occurred in her waters. The Arctic Council report claims that there is 'substantial probability' of more accidental oil spills in the Barents Sea mining areas and along the shipping routes in the Barents and White Seas.[36] Russia is poorly equipped to deal with marine oil spills.

Since there are problems with the handling of drilling spills and wastes on land, how much more difficult is it in the sea? And how much more difficult would it be under ice in the Arctic, where there is darkness for a great deal of the year, extreme cold, limited time for clean-up workers to operate safely, and dangerous sea conditions? Enormous effort was put into addressing the *Exxon Valdez* spill with 11,000 personnel, 1,400 vessels and 85 aircraft involved,[37] and yet the impacts are still being felt 20 years later.

Problems in transportation arise when oil and gas are conveyed in tankers or over large land areas in the Arctic. The Trans Alaska Pipeline System built in the 1970s moves oil north to south over 1,300 kilometres. Corroded pipes have led to spills. However the largest pipeline spill in Alaska occurred in 2006 from a corroded in-field pipeline when $760m^3$ of oil was released. In Russia, the spillages from pipelines are chronic: 'According to different sources the annual number of oil leaks and spills from oil pipelines is of the order of several tens of thousands.'[38] This should be a cause of great concern given the projected future expansion of Russian mining. The problems with pipelines breaking down will be exacerbated by the melting of permafrost (frozen ground), one of the consequences of climate change. At present permafrost supports the pipelines. This breakdown in pipeline integrity is likely to lead to additional oil spills, gas line ruptures and human exposure.

There are offshore petroleum industries in the Arctic which produce an important source of air pollution with greenhouse gases: carbon dioxide (CO_2), nitrogen oxides (NO_x), methane and non-methane volatile organic compounds. The use of natural gas to generate electricity produces high levels of CO_2 and NO_x. The non-methane volatile organic compounds mainly result from off loading of oil, particularly at loading buoys on the oil fields.[39]

The health effects of oil contamination on non-human species may involve impaired reproduction, altered DNA, reduced growth and development, or death.[40] In Chapter 6, I mention the particular threat to the Pacific grey whales from mining activities off Sakhalin. There is a specific concern about Polycyclic Aromatic Hydrocarbons (PAHs) that are components of crude oil. Table 2.1 summarises the effects on non-human species from petroleum and individual PAHs.

Table 2.1   Effects of Petroleum or Individual Polycyclic Aromatic Hydrocarbons on Organisms.

Effect[a]	Plant or microbe	Invertebrate	Fish	Reptile or amphibian	Bird	Mammal[b]
			Individual organisms			
Death	X	X	X	X	X	X
Impaired reproduction	X	X	X	X	X	X
Reduced growth and development	X	X	X	X	X	
Altered rate of photosynthesis	X					
Altered DNA	X	X	X	X	X	X
Malformations			X		X	
Tumors or lesions		X	X	X		X
Cancer			X	X		X
Impaired immune system			X		X	X
Altered endocrine function			X		X	
Altered behaviour		X	X	X	X	X
Blood disorders		X	X	X	X	X
Liver and kidney disorders			X		X	X
Hypothermia					X	X
Inflammation of epithelial tissue			X	X	X	
Altered respiration or heart rate	X	X	X			
Impaired salt gland function				X	X	
Gill hyperplasia			X			
Fin erosion			X			
			Groups of organisms[c]			
Local population	X	X	X		X	X
Altered community structure	X	X	X		X	X
Biomass change	X	X	X			

a. Some effects have been observed in the wild and in the laboratory, whereas others have only been induced in laboratory experiments or are population changes estimated from measures of reproduction and survival.
b. Includes a sampling of literature involving laboratory and domestic animals.
c. Populations of microalgae, microbes, soil invertebrates, and parasitic invertebrates can increase or decrease in the presence of petroleum, whereas populations of other plants and invertebrates and populations of vertebrates decrease.

Source: P. H. Albers, 'Polycyclic Aromatic Hydrocarbons', in D. J. Hoffman, B. A. Rattner, B. A. Burton Jr, G. Allen and John Cairns Jr (eds), *Handbook of Ecotoxicology*, Boca Raton: Lewis Publishers, 2003, p. 352.

Animals don't need to be swimming in oil to be affected. Fur-bearing mammals such as polar bears and otters can ingest it when grooming. Birds may inhale oil droplets or transfer oil from the surface to their incubating eggs. Humans may also inhale harmful gases and oil compounds, or may be affected by skin contact and ingesting contaminated food and water. A water reservoir was contaminated in Russia after the large Komi pipeline leak in 1994. PAHs pose a cancer risk to humans among a range of other health effects from minor irritations to neurological damage and death.[41] After tanker spills PAHs are likely to quickly enter the food chain, and may be stored in sediments where they are slowly released into the food chain for decades.

The indigenous peoples, especially those living in subsistence communities, may be severely harmed by oil and gas activities. Oil spills can take away their food supply or access to safe water. After the *Exxon Valdez* spill, levels of psychological stress and social disruption amongst the indigenous communities rose to great heights and persisted for years.[42]

I have already mentioned several factors that show how dangerous marine oil drilling operations are in the polar regions, but there are more. An oil spill in icy waters may become trapped within the ice, affecting the marine organisms, birds and marine mammals in the area. Dispersion and weathering of the oil will be slowed down. In cold waters oil spreads more slowly than in warmer waters resulting in lower rates of dissolution, slowing down weathering. In cold temperatures there is less evaporation of surface oil than in more temperate waters, and the darkness can reduce the rate of weathering of oil. Oil doesn't readily sink, since it is denser than sea water, unless it adheres to sediment particles. This could occur in shallow water but is less likely in the deep water of the continental shelves. In more temperate seas bacterial degeneration of oil spills takes oil out of the environment but this process is greatly slowed down in the Arctic. Chemical degeneration of oil is also much slower in the Arctic than in temperate zones.[43]

With all these dangers one would hope that response teams are ready in case of a spill. This is not the case. I have already mentioned Russia's lack of preparedness but the problem is broader. The Arctic Council report notes that 'Oil spill response in the Arctic will always be a challenge due to its remoteness, the severe environmental conditions, a limited logistical infrastructure and inadequate technology for effective oil spill clean-up in Arctic conditions particularly for oil spilled in or under sea-ice.'[44]

The scientists involved in the construction of the report included 60 recommendations for politicians. These were deleted in the published report because of opposition from Sweden and the US, both members of the Arctic Council.[45] Such political interference is unconscionable when so much is as stake.

Instead of jostling to be the biggest winner in terms of access to oil and gas fields under the Arctic, countries should take heed of these threats to the environment, fish, mammals, birds and indigenous communities with precautions to ensure that the Arctic is not destroyed for future generations. It may well be that nothing short of a moratorium on undersea mining in the Arctic will ensure this outcome. When climate change impacts are acknowledged, which unfortunately they are not in the Arctic Council report, a moratorium on mining seems even more pressing.

STRESSES ON THE ARCTIC FROM CLIMATE CHANGE

The World Wildlife Fund (WWF) has produced a very comprehensive update of Arctic climate impact science since a major assessment in 2005.[46] Many climatic changes now and in the probable future compound the threats emanating from oil and gas activities. Even the warming of the ocean and the breaking up of sea ice can be seen as a danger because they have given impetus to the mining initiatives with the consequent risks of despoliation.

The WWF report argues that 'the Arctic is not only one of the places on Earth that is most vulnerable to climate change, but also a place where vulnerability is of urgent global relevance'.[47] Furthermore they provide evidence for the view that changes are already happening across most aspects of Arctic ecosystems. Some of the changes such as the breaking up of the sea ice and the Greenland ice sheet melt are occurring much faster than predicted in previous assessments such as Arctic Climate Impact Assessment[48] and the latest Intergovernmental Panel on Climate Change Report published in 2007. Climate models now in use have not taken account of this new information. Nor has it informed mitigation and conservation strategies for the Arctic or for the world.

The Arctic is warming at 'almost twice the rate of the global mean over the past few decades'.[49] One of the lesser-known effects is the impact on permafrost. The warming of permafrost has already been mentioned as an issue in the transportation of oil and gas. It is occurring in Alaska, Canada, Russia and northern Europe, with some areas going against the trend. In northern

European Russia permafrost temperature rose from 0.2° to 1.2° –1.6° C (varying from west to east) over a 20–35 year monitoring period up to 2006.[50] Yet in Siberia currently, permafrost is weakly warming or not at all. Considering the Northern Hemisphere overall, the outlook is for a 15–30 per cent decrease in near surface permafrost by 2050.[51]

When ice-rich permafrost thaws, there is surface subsidence and thermokarsts form. These ground-surface depressions can be hazardous for Arctic biota through over-saturation or drying. The thawing of permafrost can produce greenhouse gases. This occurs when organic matter within the permafrost decomposes, releasing carbon dioxide and methane into the atmosphere. This consequence may be very significant for climate globally because of the amount of greenhouse gases that could be released: 'The upper 1–25 m. of permafrost in boreal and Arctic ecosystems is estimated to contain ~750–950 gigatonnes of organic carbon ... This indicates that permafrost is a large carbon reservoir, comparable to the atmosphere which currently contains ~730 gigatonnes of carbon.'[52]

There is also current research on methane release with permafrost thaw. The thermokarst lakes that may then form 'emit methane as opposed to carbon dioxide because permafrost beneath these lakes thaws, releasing organic matter into the lake bottom which is then decomposed anaerobically'.[53] Methane is released into the atmosphere by bubbling. If this process were to occur in Siberia alone, then ten times as much methane would be released into the atmosphere than is currently there.[54] So one major change in the Arctic with warming temperature is likely to be a *further* significant increase in warming due to the release of greenhouse gases from the thaw. These atmospheric gases will have worldwide impacts by accelerating global warming, sea-level rise and effects on living systems.

In the Arctic, contamination by oil and gas activities will combine with effects of climate change in various ways. Marine birds and mammals, for instance, will be limited in their ability to adapt to changes in the environment caused by rising temperatures after exposure to endocrine-disrupting chemicals.[55] Other effects on the marine life in the Arctic from climate change include loss of food sources for some invertebrates and fish living in the depths of the ocean. These fish eat the ice algae that drifts down during the seasonal ice melt.[56] Without the ice, there would be no algae and probably no food for these bottom dwellers. In Canada, the population size of ivory gulls living along the ice-edge year round has already plummeted with only 210 birds counted in 2005.[57] Ringed seals in the Arctic are in decline and this has been attributed to changing sea-ice conditions.[58]

Polar bears that prey mainly on ringed seals are experiencing 'a decline in body condition and reproductive output in recent years'.[59] With degradation of the sea ice, polar bears face loss of habitat. Populations of bears living in the southern areas, for example around West Hudson Bay, are experiencing significant declines in body condition and fewer cubs are being born. Even conservative climate models predict a 'two-third loss of the current population by mid-century'.[60] In 2008 the US moved to list polar bears as threatened because their sea-ice habitat is melting away. Unfortunately Canada, which is home to two-thirds of the world's polar bears, did not follow this decision. Despite the listing, the Governor of Alaska, Sarah Palin, said she hoped oil and gas activities will be able to continue without interference and she held out the possibility of a state lawsuit disputing the listing.[61] Shortly after, the US Bureau of Land Management placed a moratorium on oil drilling in the wetlands in Arctic Alaska. Grey whales in the north Pacific are declining in large numbers and this has been attributed to shifting climatic conditions in their Arctic feeding grounds,[62] but there are other threats too, as noted in Chapter 6 below.

Rural communities in the Arctic are experiencing major changes partly because of climate change. At least one community in Shishmaref, Alaska faces relocation and enormous social upheaval. This community lives on Sarichef Island off the northwest coast. The community's basic infrastructure has been undermined by 'new and extreme weather patterns, sea-ice retreat, permafrost thaw and sea-level rise'.[63]

In summary, climate change is already affecting the Arctic. Oil and gas activities could play an important role in amplifying this effect by contamination of ecosystems. There are also the greenhouse gases generated at different points of the exploration/extraction process that contribute to the global problem. When oil and gas are used in production, large quantities of greenhouse gases are released. So the continuation or expansion of these industries made possible by mining the Arctic would be a significant contributor to further global warming. In addition, rising CO_2 in the atmosphere has the potential to produce a global catastrophe in ways other than simply climate change.

OCEAN ACIDIFICATION

The oceans are becoming acidified from the rising levels of CO_2 in the atmosphere. The carbon dioxide is taken up by the oceans and because of its acidic nature it dissolves or weakens the calcium carbonate in the shells, bones and skeletons of most marine organisms including plankton.[64] The carbon dioxide in the water attacks the limestone formations of the hard corals and stops growth. The plankton are particularly important as they are the basis of the food web for a great deal of sea life including various fish populations and krill which are the main food sources for several whale species.

Plankton are also among the major producers of the Earth's breathable oxygen. If they die off then the depleted oxygen in the seas could kill marine life such as fish and shellfish.[65] As science writer Julian Cribb explains:

> a mass death of sea life combined with heavy nutrient runoff from the land could lead to stagnant, oxygen-less seas where nothing but bacteria exist. These, ironically, are the very conditions thought to have given rise to today's mighty oil fields, such as those of the Arabian Gulf, which are one of the greatest sources of the CO_2 we are pumping into the atmosphere.[66]

The plankton play an important role in absorbing the CO_2 from the atmosphere, so if they die then the ability of the oceans to absorb CO_2 will diminish. The oceans are currently the main reservoir for CO_2. If the amount of CO_2 increases and remains in the atmosphere this could trigger runaway greenhouse effects: rising temperatures that will no longer sustain life on earth.[67]

CO_2 dissolves more readily in cold water so the first line of research into acidification has taken place in Antarctic waters in order to detect whether there may be a forewarning of problems of acidification elsewhere. From Dr Will Howard's work in the Southern Ocean it has been shown that acidification is already having an affect on plankton. They are making 'thinner smaller shells than they would have in the pre-industrial ocean'.[68] Other contemporary research on coral cores from the University of Queensland in Australia shows a steady drop in calcification.[69]

This is a recently researched problem. In the 1960s the level of CO_2 in the atmosphere was 305 parts per million (ppm) and in 2007 it was 385 ppm. Until mid 2008 it was thought that the critical level for harmful effects on the ocean was 500 ppm which could be reached if changes are not made to emissions within 40 years. However research later in 2008 revealed that levels of CO_2 at 450 ppm could trigger the dangers and these levels could be reached within 30 years.[70] A recent UN report states that 'Business-as-usual climate change in the twenty-first century could make the oceans more acidic over the next few centuries than they have been at any time for 300 million years, with one exception: a single catastrophic episode that occurred 55 million years ago.'[71] That episode of ocean acidification caused mass extinction of sea creatures and the acidity levels didn't recover for 100,000 years.

Ocean acidification is, then, a global problem that will interact with climate change. Temperature changes in the oceans as a consequence of global warming are likely to compound the effects of chemical changes. Higher sea temperatures can impact on coral, causing bleaching and death especially in corals already weakened by acidification. The melting of the ice caps will lead to destruction of the habitat for krill, an important part of the food web that will already be threatened if the ocean becomes more acidic. Climate change is predicted to increase wind strength in certain areas of the ocean. This could release CO_2 currently stored in the ocean depths through the generation of upwellings and thus make the surface area of the ocean more acidic.

The exploration and exploitation of hydrocarbons is likely to degrade the local environment. The pursuit of technologies based on these energy resources is likely to degrade the world. The mining and consequent technologies are not just of concern to the coastal states, or even the surrounding regions; because they feed atmospheric pollution this is a concern for all.

Ocean acidification, and the problems of global warming and climate change, need to be addressed in order to protect the oceans. However, these are problems of a much broader scope than can be addressed in this book, requiring international co-operation to stem emissions and lifestyle changes for millions of humans. Because the consequences of letting ocean acidification build are so extreme, leading to the possibility of a runaway greenhouse effect making the world uninhabitable, it is to be hoped that international agreements will be instituted to save the planet from this fate. In this book, I am assuming that the world will remain habitable for the foreseeable future while recognising that this might be false optimism.

It is a sobering thought that whoever rules the Arctic may well influence the destiny of the world.

DIFFERENT WAYS OF VALUING THE POLAR REGIONS

There are other ways of looking at the polar regions, apart from how we can exploit their resources. The explorers Scott, Shackleton and Amundsen were fired by the ambition to reach a Pole and plant a flag. Future 'uses' of the areas seemed far from their minds, but conquest of these regions was important whatever it was taken to mean. The Norwegians Fridtjof Nansen and Hjalmar Johansen, along with Captain Otto Sverdrup, other crew and many dogs, took the sturdy ship the *Fram* into the Arctic Ocean for three years from 1893. Nansen and Johansen left the *Fram* for a 15-month excursion on the ice with a sledge and dogs. In the waters they used two kayaks lashed together with a sail (which was also used on the sledge). Nansen's journal, published in two large volumes in 1898, dwells on the explorers' relationship to the environment and their scientific findings.[72] Although they were aiming for the North Pole, the fact that they didn't reach it is given almost no importance in Nansen's writings. He comes to realise that reaching the Pole would be a matter of vanity and writes: 'Love truth more, and victory less'.[73] The focus is on the joy of finding out things about unknown places, the properties of the ice, the salinity and temperature of the ocean at different depths, the nature of the Aurora Borealis, the microscopic examination of minute undersea life forms, the behaviour of animals (including the fox who stole their sail), the detection of currents, the functioning of the *Fram* under ice pressure and so on. Along with the science, Nansen offers a stunningly evocative description of the beauty of the landscape and of his rapture in the evenings. He writes: 'How often when least thinking of it, do I find myself pause, spellbound by the marvellous hues which evening wears. The ice-hills steeped in bluish-violet shadows, against the orange-tinted sky, illumined by the glow of the setting sun, form as it were a strange colour-poem, imprinting an ineffaceable picture on the soul.'[74] Nansen's observations played a crucial role in the development of polar science, but his journals offer something in addition. He depicts the Arctic Ocean as not something to be ruled and exploited but as an area to be appreciated for the challenge it poses to human ingenuity simply to survive there, and also for the flights of imagination that it inspires.

Could these thoughts resonate with any contemporary values? Certainly they are close to the ideas expressed in Barry Lopez' book *Arctic Dreams*. Lopez recently spent several months with the Inuit, Arctic scientists, the narwhal, the musk ox, the polar bear, the kittiwake and the ice. He too explores how profoundly affected he was by the experience. Though he doesn't analyse the harm that humans could bring about, that fear is always in the background. In a remarkable gesture, when he leaves the Arctic he bows for a long time, 'to what knows no deliberating legislature or parliament, no religion, no competing theories of economics' but rather 'an expression of allegiance with the mystery of life' before a 'tangible place on the earth that was beautiful'.[75] If we destroy such beauty for short-term economic gain, how can we answer to future generations?

Moving away from the polar oceans, where the conflict around mining is building, to more temperate seas, a very different sort of battle is being waged in the courts over who should own the undersea cultural treasures, the gold and the silver. While their extraction does not pose the sort of severe environmental dangers discussed in this chapter, the disputes are generating international conflict that is likely to increase with more discoveries.

3
Underwater Cultural Heritage

Three of Shakespeare's plays begin with dramatic accounts of shipwrecks: *The Tempest*, *Twelfth Night* and *The Comedy of Errors*. Miranda in *The Tempest* proclaims:

> Had I been any god of power, I would
> Have sunk the sea within the earth, or e'er
> It should the good ship so have swallowed, and
> The fraughting souls within her.[1]

Fortunately, nowadays it is the discoveries of wrecks more often than their creation that receive attention. With more sophisticated detection equipment the treasures of the deep are exposed. Battles then ensue over who owns these treasures. Should the finders get total benefit? Should the nation that has jurisdiction over the sea where the wreck is found gain from the discoveries, or the nation who owned the vessel? Should wrecks be regarded as part of the common heritage of mankind and receive archaeological protection?

Shipwrecks are items of cultural heritage but underwater cultural heritage is a broader category. After mentioning the scope of this category I focus on wrecks since they constitute the main battleground in the present day. The issues are relatively new, sparked by the rise in importance in marine archaeology and the recent discoveries, but they need to be thought out quickly before valuable wrecks and other objects/artefacts are destroyed or their bounty lost.

WHAT IS UNDERWATER CULTURAL HERITAGE?

Underwater cultural heritage covers traces of human existence usually in a physical sense but it can also include natural objects that have been ascribed special meanings by humans possibly based on religious beliefs. The traces are usually over 100 years old and exclude cables and pipelines on the sea floor. Submerged structures, objects and paintings made by humans are included. Apart from shipwrecks, aircraft are sometimes found underwater and railway carriages sunk during transportation. Wooden, metal and stone anchors are common, and human artefacts made of gold, silver or pottery are the most sought after items, but many others contain archaeological value such as the murex or snail shell used to make a purple dye. Fish traps, especially those with a very ancient origin such as have been discovered off the coast of Britain and Australia, constitute heritage items. Miraculously, some cave paintings have survived under the sea such as those from the Palaeolithic age near Marseilles.[2] Ruins of jetties and wharves, and even harbours, exist underwater. One major harbour was discovered at Alexandria[3] and a smaller one at Calabria.[4]

Some ancient wooden round-houses have been located in waters off Britain. From an archaeological point of view, submerged villages are most interesting for their potential to tell us about early human history. The partly submerged village of Korphos-Kalamianos, 60 miles southwest of Athens, built between 1400 and 1200 BC, is currently being excavated by a joint US/Greek team.[5] A submerged coastal settlement at Aperlae in Turkey was discovered recently, which had been abandoned in the seventh century.[6]

Indigenous Australians talk about a range of special places under the sea such as a sacred hole that is the secret subterranean entrance to a spiritual domain,[7] and sacred rocks in Blue Mud Bay guarded by surrounding turbulent waters.[8] There are many sacred rocks around northern Australia, and traditional beliefs hold that no women and children are allowed near some of these sites.[9]

Shipwrecks form the bulk of heritage finds. The remains of the earliest boats made from skin, bark or reeds, or of dugout canoes, have generally been found in peat and mud beds and are unlikely to attract treasure hunters or ownership disputes. Log dugouts were uncovered in peat bogs in Denmark and are dated back to 3310 BC. In Britain dugouts from 4000–3000 BC have been discovered on land in Locharbriggs, Appleby, Short Ferry and Brigg. One of the oldest underwater finds is a dugout craft from Poole Harbour in Dorset dated to around 300–200 BC. More recent remains of dugout canoes have been found in a wide range of countries in the Northern and Southern Hemispheres. Four-thousand-year-old planked boats were unearthed in a pyramid in Egypt, but they may have been funerary boats rather than seaworthy craft. English plank boats were excavated from mud in north Lincolnshire and Yorkshire dating at 1217–650 BC.[10]

The oldest known ship was found in the Great Pyramid at Giza in Egypt, dated to about 2650 BC.[11] The oldest ship discovered underwater is the wreck of the *Uluburun*, found off the Turkish coast, dated at 1320–1295 BC. Archaeologists from the US and Turkey are still excavating this site and recovering artefacts. The list of recovered items is extensive and it is noted here as an indication of what can be retrieved from wrecks under the sea, even if the artefacts have lain there for thousands of years:

Copper and tin ingots, tin vessel fragments; whole ceramic Canaanite jars, lamps, bowls, pilgrim flasks, and juglets; coarseware and fineware ceramic shards; worked and unworked bone, shell, ostrich eggshell fragments, tortoise carapace, and ivory, the last including both hippopotamus and elephant tusks, as well as many delicately carved objects; bronzes such as bowls and bowl fragments, tools, balance pans, cauldron handles, pins, and blades; whole and fragmentary glass ingots; zoomorphic and geometric pan-balance weights of stone, lead, and bronze; moulded faience vessel shards; wood fragments; lead and tin-alloy jewellery; amber and stone beads; and stone tools.[12]

Ancient remains of other commercial ships in the Mediterranean abound, but there are no finds of the oared war ships or pirate ships mentioned in Chapter 1. They may have been pulled ashore and rotted there or, not having freight to protect them, succumbed to the gribble, the ship-eating worm that attacks timbers in salt water.[13]

SALVAGE LAWS

Marine salvage law dates back to Roman law and encapsulates the idea that if a ship and its contents are rescued from peril by a salvor (who is not the owner) then the salvor should be entitled to fair compensation. While there is broad international agreement on this basic principle there are some variations of detail in the laws of different nations. Disagreement exists over whether salvage laws should apply to ancient wrecks or other items of underwater cultural heritage.

In Greece a range of laws other than the law of salvage provide protection to ancient shipwrecks found in territorial waters. Shipwrecks of more than 50 years are classified as monuments by a ministerial decision and receive legal protection.[14]

In Britain the Merchant Shipping Act (1995) applies to shipwrecks. This Act 'includes jetsam, flotsam, lagan and derelict found in or on the shores of the sea':

> Flotsam, is when a ship is sunk or otherwise perished, and the goods float on the sea. Jetsam, is when the ship is in danger of being sunk, and to lighten the ship the goods are cast into the sea, and afterwards, notwithstanding, the ship perishes. Lagan ... is when the goods which are so cast into the sea, and afterwards the ship perishes, and such goods are so heavy that they sink to the bottom.[15]

'Derelict' means the physically abandoned vessel.[16] (Such physical abandonment doesn't take away ownership rights.) The receiver of wrecks, based in the Maritime and Coastguard Agency in Southampton, is to be notified about any wrecks in UK waters up to 12 nautical miles from the coast. If a claim to ownership of the wreck is established within one year from the time the wreck comes into the receiver's possession, then, after paying salvage fees, the owner is entitled to it. Dromgoole states that 'the lawful successors of the original owner may claim their property centuries later and the hulls of several historic wrecks in UK waters have been claimed by foreign governments'.[17] Cargo may be returned too. Seven marble sculptures dating from the second century, found in British waters, were recently returned to Turkey, the home of the original owners.[18]

If a wreck found in British waters is unclaimed then the Crown has title to it. However if the right of salvage is claimed the finder will receive the proceeds of the sale, less the receiver's expenses.[19] The strength of the application of salvage law in the UK is illustrated by a discovery off Dover. Volunteers from a diving club worked with the British Museum and the National Maritime Museum, who paid for professional archaeological work, to recover bronzes from an ancient wreck site. As the finders of the site, the volunteers were technically the salvors and were awarded the full purchase price of the 300 bronzes they found, less the receiver's fee, even though the expedition had been publicly funded.[20]

The term 'treasure salvor' is enjoying some currency as there is no clear distinction between a treasure hunter and a salvor where heritage items are concerned. Both usually go after items of high financial value and thanks to the existence of salvage laws they have been able to carry out their operations legally. Where nations try to make ownership claims on underwater heritage in their waters, they may reject the right of treasure salvors to operate there, even going so far as labelling such salvors as 'pirates'.

A report on a discovery by Chinese archaeologists of an ancient sunken ship containing Ming Dynasty porcelain in the South China Sea states: 'it looks like hi-tech pirates got to it before them. ... [The] police confiscated 21 pieces of porcelain from a fishing boat whose owner claimed that divers he had hired for deep-sea fishing had recovered the porcelain by accident.'[21] While it may be a strange use of the term 'pirate' for treasure salvors, there are reports that the well-organised Somali piracy network is involved in trafficking in archaeological remains.[22] Since dissatisfaction is growing with the application of the law of salvage to underwater cultural heritage, it is timely to ask who should get ownership of these discoveries.

TREASURE SALVORS AND OWNERSHIP

Should the treasure salvors get total benefit? Marine exploration requires a high level of financial commitment, to buy or lease a boat and equipment, which may include scuba gear, robots, computers, remotely operated vehicles, depth-finding equipment, underwater cameras, underwater sonar and magnetic devices. There is also much skill involved in using the equipment or diving over wrecks, and also in the museum or library searches undertaken beforehand, as well as the fund raising. Bad weather conditions, entanglement of scuba equipment, other diving accidents or equipment failures are constant concerns. The quest may fail to find anything of value.

However, if treasure salvors expend a lot of money, time and risk in trying to find treasure on land they are not in the present day usually rewarded with the full value of their finds, at least if they are acting within the law. So why should it be different in the oceans? It could be argued that if treasure salvors are not legally entitled to their finds there would be no incentive to explore. This is true. However if the results of the exploration are secured by private individuals and sold to private individuals perhaps it is better to leave those materials in the oceans until an ownership policy is accepted that promotes a greater benefit. If the treasure salvors sell to museums then some public benefit does follow. However there is a danger of corrupted archaeology if the financial cost and benefit are the prime factors. For instance, shipwrecks may be blasted in an attempt to get the marketable items out, and important associations between marketable and non-marketable items may be lost, or collections split up. These outcomes degrade the cultural heritage value of the find. Gould argues that 'treasure hunters have contributed to or created underwater chaos by planting materials from other sources'.[23] There is a growing tendency for treasure hunters to appropriate archaeology as a means of enhancing the monetary value of finds,[24] and deception is sometimes used to invent an archaeological story around a shipwreck.[25]

There are institutions propping up salvors' activities such as shipwreck fairs, art markets in major Chinese cities, Christie's in Amsterdam and various museums. Even so, no compelling reasons exist for granting treasure salvors the total benefit of their finds. This holds whether the retrieval is legally sanctioned or not. It is a moral issue about what should happen to heritage items and in this area the law lags behind contemporary moral sentiment.

NATIONAL OWNERSHIP

Should the nation that has jurisdiction over the sea where the wreck is found gain from the discoveries, or the nation who originally owned the vessel? Since the Law of the Sea allows coastal states to claim 200 nautical miles from their coast as an EEZ, it would seem to follow that such states should have ownership rights over cultural heritage items in those waters. The Law of the Sea is in fact confusing and contradictory on this issue. According to Article 303, part 1: 'States have the duty to protect objects of an archaeological and historical nature found at sea and shall cooperate for this purpose.'[26] The expression 'at sea' is very vague. It is unclear whether the duty applies to the whole EEZ or only a part. The second section of Article 303 refers back to Article 33 which defines a contiguous zone as 24 nautical miles from the coast and says that a coastal state may exercise the control necessary to 'prevent infringement of its customs, fiscal, immigration or sanitary laws and regulations' within this 24 mile zone. Then Article 303 states that: 'In order to control traffic in such objects [archaeological or historical], the coastal State may, in applying article 33, presume that their removal from the seabed in the zone referred to in that article without its approval would result in an infringement within its territory or territorial sea of the laws and regulations referred to in that article.' However this requirement does not affect the rights of identifiable owners, the law of salvage or other rules of admiralty, according to section 3 of Article 303. Herein lies a major problem. If the law of salvage applies then states may well fail in their duty to protect objects of archaeological significance.

Several states that have signed the Law of the Sea now have legislation covering archaeological heritage within the 24 mile zone. These include France, Tunisia, China and Denmark.[27] Spain, Portugal, Ireland, Australia and Jamaica have expanded the protection further to take in the EEZ.[28] In the US, however, where treasure hunting is a thriving business supported by strong lobby groups such as ProSEA, there is very little protection of wrecks or any other underwater heritage. The Abandoned Shipwreck Act of 1987, interpreted on a state-by-state basis, gives cover to some abandoned historic shipwrecks but only within three nautical miles of the coast (or nine nautical miles in the Gulf of Mexico). Under this Act, states are given title to certain abandoned historic shipwrecks and they can make laws regarding their management. Wrecks receive variable preservation coverage. The law allows commercial salvage of wrecks. Preservationists along with treasure salvors are regarded as interested parties and their interests are to be considered in the use of shipwrecks.[29] As mentioned above, the US is not a signatory to the Law of the Sea.

To legally salvage an abandoned ship in the US EEZ outside of the three (or nine) mile zone, a salvor may apply to the admiralty courts under the law of finds. The finder can then be given possession of the wreck to exploit how they wish. Alternatively if the salvage law is used, a salvor is often awarded the vast majority of the recovered treasure.[30] Ella argues that treasure hunting is even encouraged beyond the three (or nine) mile zone, which is clearly against the duty to protect objects of 'an archaeological and historic nature'.[31] There have been some attempts in the US courts to overturn decisions based on the law of finds or salvage law but they have been mostly unsuccessful. The Federal Government took on the treasure salvors of the Spanish galleon *Neustra Señora de Atocha*, which sank in 1622 well beyond the three mile zone off Florida. The government was acting to protect the wreck. They lost and *Atocha* is still being privately exploited. Gold, silver, emeralds, artefacts and armaments valued at more than $400 million in 2009 have been recovered by Mel Fisher and his company. The Fisher family has the rights to all this treasure. The extraction process is crude: 'Before a dive begins a hole is blown in the sand using thrust from propellers on salvage boats redirected downwards. What's left is a 15 foot crater filled with limestone, shells, an occasional stingray and history.'[32] Divers then go down and retrieve any 'history' they deem of value. In 2009 *Atocha* coins were available on the internet thus ensuring wide dispersal.

One very important document abetting the plunder of Spanish wrecks was a letter written by Marquis de Morry del Val, the Spanish Ambassador to the US in 1965, concerning Spanish wrecks off Florida, in which he said:

> there is no doubt that the Spanish State may not claim any title to said treasure for the following reasons:
>
> 1. If the discovery is considered 'marine salvage', the owner of the ship and/or merchandise would have lost all rights because he abandoned any attempt of recovery.
> 2. If the discovery is considered a discovery of a treasure in the territory ... under the jurisdiction of a state ... the laws of this state will determine title to the treasure.[33]

After 35 years, Spain has rethought this position, especially after the US Supreme Court decided that abandonment of a wreck has to be explicit, not just implied, for the purposes of working out possession. In 2006 a US Court of Appeal awarded Spain ownership of the wrecks of the Spanish Royal Naval ships *La Galga* and *Juno* that sank in 1750 and 1802 off the coast of Virginia. Many artefacts and coins were retrieved from these wrecks and the Spanish Embassy agreed to place them on public exhibition at Assateague Island National Seashore, close to where the vessels were lost.[34] This was an important decision, received with great concern in the treasure salvor community who have labelled it 'un-American'.[35] It was, however, a victory for those who would like to see historic sites commemorated and the artefacts and coins available for public appreciation.

Spain hasn't been so successful in other ownership claims on wrecks. The Spanish galleon *La Capitana Jesus Maria*, which sank off Ecuador in 1654 carrying gold, silver and precious stones, was discovered by a Norwegian salvage team in 1997. The Ecuadorian Government agreed to split the very significant proceeds of the wreck with the Norwegian team.[36]

Other nations with seafaring histories, such as Holland, have also been concerned about the ownership of their wrecks. The Dutch East India Company went bankrupt in 1798 and was taken over by the Batavian Republic that became the Kingdom of the Netherlands in 1813. According to Gould, the Dutch Government claims ownership of all the company's wrecks wherever they are.[37] However, there is a Dutch/Australian agreement whereby Holland gives up rights to wrecks of the Company's ships off Western Australia so that the archaeological collection can be kept together – it is held in Australia with some duplicates in Holland.[38] Germany also makes no claims on the wreck of its warship, the *Kormoran*, found in 2008 in Western Australian waters. Australia is not seeking ownership of the wreck of the submarine AE2 found in 2008, in Turkish waters. Of course it is easier to resist ownership claims when gold and silver are not at stake.

If a coastal state has particularly treacherous waters or hazardous weather patterns a principle which assigns the state ownership of all wrecks off the coast seems unjust. It could be called a spider-web basis for ownership. However, if the country of origin is deemed to be the legal owner of a wreck then this could prompt a development of marine exploration into the waters of foreign countries. If their quest is legally legitimate then coastal states should not have the power to block exploration and retrieval in their waters. However these ventures are likely to clash with what a nation can and cannot do in coastal waters of a foreign state as laid down in the Law of the Sea (see Chapter 1 above.)

In summary, the policy of not regarding physical abandonment of wrecks as the crucial factor in determining ownership is a fair one. If the original owner is clear and they have not expressly abandoned the wreck then they should have a greater claim on the wreck than the state that has jurisdiction over the waters where it is found. The problem of access resulting from the Law of the Sea regulations means either that the Law should be changed or that a different ownership solution is needed. Also some policy for wrecks with no known original owners is necessary.

COMMON HERITAGE

Should wrecks be regarded as part of the human common heritage? The destruction of a shipwreck is forever. Yet shipwrecks are resources for finding out about human history and prehistory sometimes in a way impossible on land. Gould claims that underwater environments may 'preserve complex associations of cultural remains better than they can be preserved on land. Many important issues in human prehistory may ultimately be resolved by archaeology underwater.'[39] The heritage is of value not just to the coastal state or state of origin but to many people living elsewhere and to future generations, suggesting that underwater cultural heritage should be placed under the control of an international body. However this also amounts to a challenge to the Law of the Sea that awards coastal states sovereign rights and jurisdiction up to 200 nautical miles. The UNESCO Convention on the Protection of Underwater Cultural Heritage, adopted in 2001, tries to get around this issue, but not entirely successfully. The Convention acknowledges 'the importance of underwater cultural heritage as an integral part of the cultural heritage of humanity',[40] and expresses deep concern at the increasing commercial exploitation of such heritage, in particular the activities aimed at sale, acquisition or barter. The Convention is 'committed to improving the effectiveness of measures at international, regional and national levels for the preservation *in situ* or, if necessary for scientific or protective purposes, the careful recovery of underwater cultural heritage'.[41] State Parties to the Convention are required 'to preserve underwater cultural heritage for the benefit of humanity'.[42] They should use 'the best practicable means at their disposal and in accordance with their capabilities'.[43] The Convention supports 'responsible non-intrusive access' to heritage sites and respect for all human remains underwater.

National sovereignty and jurisdiction remain unchanged. The Law of the Sea is held to be consistent with the Convention. It is not envisaged that there will be an international presence in the waters of coastal states. Rather, in signing the Convention coastal states are agreeing to preserve heritage in their waters. Salvage law can still apply to recent finds or boats in peril. However there are obstacles in the way of applying it to heritage items especially as commercial exploitation is disallowed. The rules of the Convention are supposed to cover all of the sea but the high seas require a different response more amenable to international input. Heritage discoveries in the high seas are to be reported to the Director General of the UN and the International Seabed Authority. Any state can then declare an interest in how best to preserve the heritage, particularly if there is a cultural, historical or archaeological link to that state.

The Convention is an impressive document but its passage has been slow. Twenty ratifications were needed to bring it into force and these were achieved in 2008. The Convention then entered into force on January 2, 2009 for those 20 states and for any other state three months after their ratification. It is mainly small states without a seafaring history that have ratified the Convention, except for Spain and Portugal. One of the ratifying states, Panama, offers flags of convenience. These flags are a negative force in the oceans. They assist pirates, illegal fishers, polluters and various other criminal activities such as arms trafficking. Traditionally a ship was registered in the shipowners' national state. Shipowners were then subject to the laws of this state when at sea and they were entitled to fly the national flag. It is now possible and very common to register with a country that offers a flag of convenience. On payment of a fee, the shipowner may register with any one of 40 countries presently offering these flags, including the land-locked Mongolia. The advantage for the shipowner is that labour laws and environmental standards are weaker than in many other countries and surveillance is poor.[44] The impetus for this practice is economic and neither the Law of the Sea nor other international law offers any impediment.

Neither the US nor the UK have ratified the Convention on the Protection of Underwater Cultural Heritage, yet they are the best equipped for underwater exploration. Their stand will make it very difficult to enforce the Convention in international waters. Treasure salvors in the US gleefully declare the Convention to be a dead issue.[45]

One current conflict well illustrates the urgency of the need for decisive action to protect heritage. It concerns the battle, mentioned in the Introduction, between a US salvage company called Odyssey and the Spanish Government over the discovery of the wreck of a Spanish galleon, now thought to be *Nuestra Señora de las Mercedes*, that sank off the Portuguese coast in 1804. The wreck contained Spanish coins with an estimated value of $747 million in 2007.[46] These coins were taken to Florida from Gibraltar. One of the search vessels, *Odyssey Explorer* was detained in October 2007 in the Spanish port of Algeciras, and the captain arrested for not allowing a search of the vessel. He was released and the ship was allowed to depart after four days, when it became clear that the treasure was not on board. The *Mercedes* is of value not only because of the coins but because she was part of an important episode in human history: the Spanish Armada. Odyssey claims that they will be entitled to 90 per cent of the value of the recovery under US admiralty law, even if it is accepted that the wreck is Spanish and Spain has not expressly abandoned it. They claim to be acting in conformity with US salvage law and the Law of the Sea.[47] This case is being heard in Florida where Spanish lawyers are arguing that both the wreck and the treasure are theirs. The Spanish Director of Fine Arts said that 'the action of Odyssey in the case had been "morally and legally unacceptable"', adding that 'historical reason and legal sufficiency were on the side of the Spanish state [and that he had] the moral conviction that Spain would win the case'.[48] In May 2008, the Spanish Government demanded that Odyssey hand over more than

$530 million worth of the salvaged treasure.[49] It is anticipated that this case will not be resolved until 2010.

In the case of the wreck of the *Titanic*, heard before the Admiralty Court in Virginia, the decision was to award salvage rights to the salvors who were American and French. The *Titanic* was a British ship and the wreck was found in international waters, yet the US decision held in the face of opposition from the French Institute for Maritime Research and Exploration. Subsequently the US drafted an agreement with the UK, Canada and France that would make the *Titanic* a memorial *in situ* with severe limitations on commercial use. However by this time 1,800 items had already been removed.[50] By mid 2009 the US and the UK had signed the agreement.[51] The *Titanic* sank in 1912, too recently to normally be considered 'heritage', but it is a key item in the western public imagination and deserves special treatment for that reason.

The Law of the Sea does have an article covering all objects of an archaeological and historical nature found in the high seas and it states that they 'shall be preserved or disposed of for the benefit of mankind as a whole, particular regard being paid to the preferential rights of the State or country of origin, or the State of cultural origin, or the State of historical and archaeological origin'.[52] It is clear that this aspect of the Law has been ineffective. Nevertheless it is ironic that important decisions about the treatment of shipwrecks have been made through US courts while the US is not a signatory to the Law of the Sea. Whether or not these decisions favour the salvors or the preservationists, if mankind as a whole is to benefit it would seem fairer for decisions to be made in an international court rather than the court of one state.

The Convention offers a means of clarifying a very murky present situation where salvage law can block genuine national desires for preservation of heritage within the waters of coastal states. If a states' heritage is lying in the waters of another state there are often huge impediments to preservation of wrecks, as illustrated above, and the situation on the high seas is fraught.

The UK has refused to ratify the Convention mainly because of its opposition to the 'blanket' approach to the protection of archaeological sites within British territorial waters.[53] At the Burlington House Seminar in 2005 British archaeologists suggested ways around this which have since been ignored, leading to the suggestion that the UK has no political will 'to actually [do] something to protect underwater cultural heritage internationally'.[54] It is noteworthy that in 2003 the British Ministry of Defence entered into a joint excavation venture with Odyssey to investigate the wreck of the warship the *Sussex*, lying east of Gibraltar. The ship was lost in a storm in 1694, carrying gold coins sent by the English to buy the Duke of Savoy's loyalty in the nine years' war. There is an estimated $8.2 billion worth of treasure with this wreck, and the UK has agreed to share the proceeds with Odyssey.[55] In an article in 2007, Ian Jack reported that this agreement 'dismayed marine archaeologists'.[56]

The US Department of State expressed concern over the extension of state control over underwater cultural heritage to the limit of the EEZ,[57] arguing that it would be some time before many countries would be capable of exercising such control. This seems a lame excuse for refusing to ratify the Convention, given that it takes account of varying capabilities. A more likely reason for not ratifying is that it protects the interests of treasure salvors.

In discussions around the Convention the sticking points seem to relate to matters within coastal boundaries more than in the high seas. The UN has chosen not to interfere with the ocean boundaries laid down in the Law of the Sea, but it may be necessary to do so to be able to move forward on this issue. The Convention claims to be consistent with the Law of the Sea, and it is so for the high seas. However the Law has a different approach to waters up to 200 nautical miles out from the coast, as indicated above, and although it is unclear and inconsistent it allows salvage law to operate in a way prohibited in the Convention. If the waters beyond the 12 mile zone were regarded as high seas and under the control of an international body, then this would go some way to addressing these problems. If the Convention was then adopted for the re-drawn high seas, the actions which would flow from it could have a persuasive effect on coastal states and protect underwater heritage for humanity.

Without radical changes in ocean governance international conflicts over undersea resources and cultural heritage are likely to develop. We are only now seeing the faint beginnings. Meanwhile, on the surface of the oceans, more obvious battles are being waged between pirates and those who go after them.

4
Modern Piracy and Terrorism on the Sea

At sea anything can happen.

> Mintodo (a convicted pirate), *The Outlaw Sea*[1]

Every generation gets the pirates it deserves.

> Robert I. Burns, *Muslims, Christians and Jews*
> *in the Crusader Kingdom of Valencia*[2]

THE *ALONDRA RAINBOW*

At 10.30 p.m. on October 21, 1999, Captain Ikeno was writing up a departure telex on the *Alondra Rainbow* when he heard some thumping overhead and shouts on the intercom. His ship had been taken over by pirates. The *Alondra Rainbow* was a new 370-foot cargo vessel owned by a Japanese company but registered in Panama. The crew was Filipino except for the Japanese captain and chief engineer. They had just picked up a shipment of aluminium ingots valued at $10,000,000 from Kuala Tanjung in Sumatra, at roughly the midpoint of the Straits of Malacca. The shipment was headed for Omuta in Japan. The sea was calm in the Straits. The moon was full. The crew were tired. At 10.00 p.m. Captain Ikeno switched on the autopilot and went to his quarters, reminding the third officer to keep a look out for pirates.[3]

Over the previous days the pirates had been assembling on the *Sanho*, an old freighter turned pirate ship anchored off Jakarta. This was a big operation. There were 35 people involved, mainly Indonesians but also Chinese, Malaysians and Thais. Only 15 took part in the actual hijacking. They used a small fast boat launched from the *Sanho* and entered the *Alondra Rainbow* from the stern.[4]

After the disturbance Captain Ikeno rushed to the bridge and was overpowered by men armed with knives, guns and swords. The pirates tied the captain's hands behind his back and forced him to guide them to the other crew. They were bound and blindfolded and herded into the crew's mess room with threats of death for any rebellion. The pirates steered the *Alondra Rainbow* through the Straits for two hours. Then the crew were taken onto the deck.

The *Sanho* pulled up alongside with many armed men aboard. The crew were forced onto the *Sanho* where they stayed for nearly a week. They were fed only twice and given dirty drinking water from an ESSO can. The *Alondra Rainbow* had been taken away by some of the pirates after the crew transfer. On the seventh night the crew were forced into a life raft the pirates had taken from the *Alondra* and cast adrift. The *Sanho* steamed off into the night. They drifted for ten days with minimal survival rations. Passing ships ignored their flares. On the tenth day a small commercial fishing boat approached. The Thai captain was suspicious and wanted to see passports which they no longer had. Captain Ikeno was eventually able to write the name '*Alondra Rainbow*' and showed this to the Thai captain who radioed for information and reassured, took the crew on board. The next day they were put ashore at Phuket amongst the holiday crowd. Captain Ikeno and the chief engineer flew home to Tokyo never to return to sea.[5]

The *Alondra Rainbow* was hijacked ten years ago when Indonesia and the Straits of Malacca formed the hub of pirate activity. In 2000 there were 119 actual or attempted piracy attacks in Indonesia and a further 75 in the Straits of Malacca that divide Indonesia from Malaysia. There were 469 actual or attempted attacks globally for that year.[6] From January to September 2009 there were two reported piracy incidents in the Straits of Malacca and seven in Indonesia out of 306 globally.[7]

Ten years ago there was some activity off Somalia with 14 actual or attempted piracy raids but no reported incidents in the Gulf of Aden/Red Sea.[8] Northern Somalia borders the Gulf of Aden. Piracy in the Gulf over the next few years peaked at 18 incidents in 2003 then declined, until rising to 92 in 2008 and 115 for January to September, 2009.[9] Off eastern Somalia piracy incidents declined to only two in 2004, peaked at 35 in 2005 and declined again to 19 in 2008.[10] From January to September 2009, 47 incidents were recorded.[11] Somali pirates were also responsible for six other attacks in more distant waters prior to October in 2009 and several attacks near the Seychelles in November, 2009.[12] As we can see from these figures, contemporary piracy should not be viewed as a phenomenon affecting world shipping only in 2009 and only off Somalia.

THE LAW OF THE SEA AND CONTEMPORARY PIRACY

The Law of the Sea defines piracy as an activity that occurs exclusively on the high seas. Coastal states have full jurisdiction in their territorial sea and qualified jurisdiction in the rest of the EEZ. Other states can freely navigate the waters of this area but according to Article 58 (3) of the Law of the Sea: 'In exercising their rights [such as the right of navigation] ... under the Convention in the exclusive economic zone, States shall have due regard to the rights and duties of the coastal State and shall comply with the laws and regulations adopted by the coastal State.'

Restricting piracy to the high seas even if pursuit is allowed into the EEZ has given rise to confusion and delays in preventative actions, as I will illustrate below. Most of the piracy attacks just listed, including the hijacking of the *Alondra Rainbow*, would not count as piracy as the ships were boarded within EEZs. This is the case with Somali raids also except for recent hijackings in the high seas. The full definition of piracy in the Law of the Sea is as follows:

(a) any illegal acts of violence or detention, or of any act of depredation, committed for private ends by the crew or passengers of a private ship ... and directed:
   (i) on the high seas, against another ship ... or against the persons or property on board such ship;
   (ii) against any ship ... person or property in a place outside the jurisdiction of any State;
(b) any act of voluntary participation in the operation of a ship ... with knowledge of facts making it a pirate ship;
(c) any act of inciting or intentionally facilitating an act described in sub-paragraph (a) or (b).[13]

The Straits of Malacca are very narrow, at some points only 1.5 nautical miles wide. Although up to 300 nautical miles wide in some places, the sea territory still falls within the EEZs of the bordering states of Indonesia or Malaysia. There are no high seas in the Straits. Yet the Law of the Sea states that piracy only occurs on the high seas. The *Alondra Rainbow* was taken over in seas that are under the jurisdiction of Indonesia. Hence the crime is armed robbery at sea which would be covered by the Indonesian criminal code. This is not a pedantic question about what the crime is called. It concerns who has jurisdiction over the area and whether the crime should fall under national or international law. If a crime at sea is deemed to be a pirate attack then according to the Law of the Sea 'every State may seize a ... ship taken by piracy and under the control of pirates, and arrest the persons and seize the property on board. The courts of the State which carried out the seizure may decide on the penalties to be imposed.'[14]

The criminals on the *Alondra Rainbow* were captured hundreds of miles away from Indonesia almost one month after the attack. The ship had been re-painted, re-named and re-registered but a captain on a passing Kuwaiti freighter, primed to look out for the *Alondra Rainbow* from the piracy-reporting centre in Kuala Lumpur, phoned in his suspicions. The ship was intercepted by the Indian navy operating under the encouragement given by the Law of the Sea to any state to apprehend pirates no matter where on the high seas their attack took place.[15] The legality of this capture is murky. The criminals did not fit the definition of 'pirate' in the Law of the Sea but India, for whatever reason, wanted to use its naval strength to capture them and bring them to court.

Before the Indian navy boarded the *Alondra Rainbow* the pirates opened sea valves to try to scuttle the ship. These were closed off by Indian naval divers. The men were taken to Mumbai for trial. This was 'a landmark case as it was the first in history to use international law and specifically the Law of the Sea to claim "universal jurisdiction" for an act of piracy having nothing to do with the prosecuting country'.[16] However, it was not a tidy use of the international law since, strictly according to that law, the criminals were not pirates; and the situation gets even messier since piracy is not assigned any penalties under Indian law. The Indian Government, keen to make a prosecution, charged the pirates with other crimes: 'armed robbery, attempted murder, assault, theft, forgery and fraud – and even of entering India without valid passports, though they had done so involuntarily'.[17] If these are the crimes the criminals committed then they should be put on trial in the country where the crimes took place. To summarise the confusion: India had a right to try the criminals if they were pirates, which they were not strictly speaking, but even if they were they were not charged with that offence but with others for which there is no universal jurisdiction. The trial took three years. The accused were found guilty of all charges except the passport violation and sentenced to seven years imprisonment. Captain Ikeno's evidence proved decisive.

The UN International Maritime Organisation (IMO) based in London, has been keeping records of pirate attacks since 1995. It also records armed robbery against ships, which is defined as 'any unlawful act of violence or detention or any act of depredation, or threat thereof, other than an act of "piracy", directed against a ship or against persons or property on board such ship, within a State's jurisdiction over such offences'.[18] These offences may occur on a ship at sea or in port.

The International Chamber of Commerce (ICC) runs the International Maritime Bureau (IMB), a piracy-reporting centre in Kuala Lumpur, Malaysia, which alerted ship captains to the hijacking of the *Alondra Rainbow*. It produces similar reports to the IMO and also an up-to-date piracy map locating high risk areas globally. The IMB groups piracy and armed robbery against ships. The two crimes together include 'any act of boarding any vessel with the intent to commit theft or other crime and with the capability to use force in furtherance of the act'.[19]

The total number of actual and attempted attacks reported to the IMB since 1994 is noted in Table 4.1.

Table 4.1 Piracy and Armed Robbery Against Ships: Actual and Attempted Attacks Since 1994.

1994	90	2002	370
1995	188	2003	445
1996	228	2004	329
1997	248	2005	276
1998	202	2006	239
1999	300	2007	263
2000	469	2008	293
2001	335	2009 (Jan–Sept)	306

Source: International Chamber of Commerce, International Maritime Bureau, 'Piracy and Armed Robbery Against Ships', Annual Reports, Jan 1–Dec 31, 2005, 2006, 2007 and 2008. Report, Jan 1–Sept 30, 2009.

The main locations are the Gulf of Aden/Red Sea, Somalia, Indonesia, Malaysia, Straits of Malacca, Vietnam, Bangladesh, India and Nigeria, with scattered attacks in dozens of other countries. All the attacks are listed by country. There is no 'high seas' category. One new feature that these figures disguise is the rise in hijackings, continuing in the present day. The IMB figures for hijacks are in Table 4.2.[20]

The Somali pirates are behind the hijack increase. The common pattern is to take a ship, sail it into a Somali port taking the crew as hostage and then demand and secure ransoms for the safe return of the ship and crew. In 2008, 889 people were taken hostage, mainly by Somalis.[21] From January to September 2009, 661 hostages were taken, 523 by Somali pirates.[22]

Table 4.2   Actual and Attempted Hijacks of Ships Since 1994.

1995	12	2003	19
1996	5	2004	11
1997	17	2005	23
1998	17	2006	14
1999	10	2007	18
2000	8	2008	49
2001	16	2009 (Jan–Sept)	34
2002	25		

Source: International Chamber of Commerce, International Maritime Bureau, 'Piracy and Armed Robbery Against Ships', Annual Reports, Jan 1–Dec 31, 2005, 2006, 2007 and 2008. Report Jan 1–Sept 30, 2009.

Another recent change is in the use of firearms. Knives used to be the favoured weapons. Now it is automatic weapons and rocket-propelled grenades. However, the number of people killed or missing in piracy attacks has dropped from 92 in 2003 to 32 in 2008[23] and 14 between January and September 2009.[24] As the mode of attack shifts to hijackings, it is in the interests of the pirates to keep the crew safe for ransoms.

While hijacks are not likely to be under-reported this is not the case for other acts of piracy. Investigations can cause costly delays and possible increases in insurance premiums. Some shipping companies actively discourage reporting incidents, or have a policy of doing so only privately, so the IMB or IMO never get to hear of these cases.

When thinking about piracy it is common to accept the broad IMB definition and not be concerned about whether or not the attack took place on the high seas. This marginalises the Law of the Sea, which was an attempt to capture Gentili's sentiments noted in Chapter 1. In the sixteenth century, however, coastal states did not have such a broad jurisdiction. Herein lies a major problem. The crime of piracy has universal jurisdiction in that all states are encouraged to bring pirates to account. The Law allows pursuit into a foreign state's EEZ though not their territorial sea. Coastal states own their territorial sea but they also have certain rights and jurisdiction over the EEZ supporting the idea that this area too is primarily a state responsibility. It may be thought that the universal jurisdiction for piracy simply overrides any of the state's governance of the EEZ, but given the lack of a central body to give substance to universal jurisdiction or equity to any prosecution this could lead to undesirable outcomes. First, there may be a lack of willingness to pursue pirates into a foreign states' EEZ because of a perceived conflict over who should be responsible for pursuing pirates in this area. Second, there may be a reluctance to take on the task of prosecuting captured pirates if it is deemed that the coastal state should be involved. Finally there could be suspicions about the motives of those pursuing the pirates. (Some Somalis currently accuse the naval forces guarding shipping corridors of protecting illegal fishing in their waters.[25])

In Chapters 2 and 3 I pointed out that the dominance of private or national interests in exploiting undersea non-living resources and heritage beyond the territorial seas is leading to international tension. I suggested that one way forward could be to re-draw ocean boundaries, fix state limits at 12 nautical miles, and place the rest, the new high seas domain, under the control of an international body. Piracy may well provide another reason to move in that direction. The IMO and the IMB have tackled the issue of how to categorise piracy and sea robbery in different ways, further highlighting problems with the Law of the Sea definition. Both combine the incident reports for piracy and armed robbery against ships. The IMO data lists whether the attack occurred in international waters, territorial waters or port areas. The IMB data only give a breakdown between 'berthed', 'anchored' and 'steaming'.

The crimes committed on ships berthed or at anchor in port tend to involve minor theft of goods or cash with little violence. Dana Dillon argues that 'adding these petty crimes to the reports of piracy both exaggerates the problem and blurs efforts to make policy'.[26] She wants the definition of piracy to include all ships underway whether on the high seas or not. In addition she recommends a change in the law to allow pursuit of pirate vessels by foreign naval air and sea vessels 'into territorial waters with notification to and approval from the relevant authorities'.[27] Alternatively she suggests regional agreements with neighbours to coordinate air and sea patrols and permit entry into the territorial waters of foreign states in pursuit of pirates.

There have been some co-ordinated measures adopted by Indonesia, Singapore and Malaysia to counter piracy in the Straits of Malacca. These patrols seem to have been successful as there has been a drastic drop in piracy in the Straits. However, it is not known how long the patrols will operate,[28] while the Indonesian navy predicts 'that with the global crisis there will be more pirates and illegal activities taking place in the Malacca Strait'.[29] Around Indonesia, where there is no joint regional effort, piracy is still a problem. A bulk carrier was boarded and cash taken by pirates on November, 16, 2009. The number of pirates involved and the size of the ship suggest that piracy is becoming more organised around Indonesia. The IMB Reports repeatedly say that there is under-reporting of piracy around Indonesia.

Although local co-operative strategies such as those in the Straits of Malacca should be encouraged they are inadequate on their own. Regional agreements may be impossible to establish because of the strained relations that often exist between neighbouring states in Asia and Africa and the conflicts generated over sea boundaries and fishing rights. There is also a definite possibility that some governments officially sanction piracy, as suggested by the London-based Control Risks Group and a variety of maritime bodies.[30] This points to the desirability of an international response.

Dillon's suggestion that the term 'piracy' should apply to all attacks on ships underway is a good one, but it returns us to the jurisdictional conflict mentioned above. The lack of a clear resolution of this conflict is highlighted in the uneven responses to the capture of Somali pirates, as discussed below.

WHY PIRACY NOW?

Piracy as a phenomenon of modernity, where attacks are made on international shipping and yachts, can be linked to the end of the Cold War, usually dated to around 1990. This led to a decrease in the size and reach of navies that were likely to have deterred pirates. Also with the end of the Cold War, maritime trade increased. Tight passages such as the Straits of Malacca and the Gulf of Aden offer commercial advantages over taking the longer way around. However, sailing these routes means that ships are close to pirate launching places and traffic build-up slows them down. Recent advances in outboard technologies mean that small craft can catch up with even very large tankers such as the *Chaumont*, a crude oil tanker 338 metres long, and the *Sirius Star*, also over 300 metres. Other new gadgets also help in the organisation of raids: mobile phones, handheld satellite navigation systems and portable ship-to-ship/shore radios.[31]

Heavy weapons have become easier to acquire, especially in Southeast Asia, sometimes from the stocks left over from conflicts in Cambodia and Afghanistan. Other weapons have made their way to this region from the former USSR and Eastern Europe, smuggled out by various crime syndicates.[32] Somali pirates obtain weapons from Yemen or buy them in the capital Mogadishu. Despite an arms embargo, weapons make their way into Mogadishu from Yemen, financed by Eritrea, who appear to be waging a proxy war with Ethiopia in Somalia. Ethiopian troops have been in Somalia off and on for some years, withdrawing in January 2009 and returning again in February. The chaotic political situation fuels the desire for arms which then can easily fall into the hands of pirates. In November 2009, they took over a United Arab Emirates cargo ship which was heading for Somalia with a shipment of weapons circumventing the UN arms embargo.[33] The fate of these weapons is presently unknown.

There is reluctance to arm the ships that are the targets of pirate attacks. If merchant ships or cruise ships carry firearms this could pose a risk to lives by escalating violence or producing explosions. Israeli security guards used pistols to deter pirates from attacking the cruise ship *Melody* in 2009. Roger Middleton, an English expert on Somali piracy in the London-based think-tank Chatham House, said this action violated the consensus view in the shipping industry that weapons should not be carried. A major impediment to carrying arms is also the Law of the Sea. As noted in Chapter 1, ships normally have the right of innocent passage through territorial waters of a coastal state. This right is premised on the belief that merely passing through poses no threat to the coastal state or others using those waters. If ships carry arms then this belief might be questioned and much tighter legal restriction concerning the right of passage could be placed on shipping.

Yachts may carry arms but there is an obligation to declare firearms to port officials. If this is not done, jail sentences can result.[34] Also, producing or using a gun can be a provocation to pirates to shoot to kill. Peter Blake, a New Zealander aboard the yacht *Seamaster* anchored in the Amazon, fired at pirates and was killed in return fire.

The motivation for pirate attacks is growing along with the valuable removable items. Even yachts are likely to have portable electronics and cash. Merchant ships often have large amounts of cash to pay crew and port fees, as well as portable electronics, medical chests, the crews' personal belongings and the cargo. The crew and the ship attract increasingly larger ransom figures, especially in Somalia. The pattern set up in 2008 was to hijack a ship and hold it with the crew, sometimes for months, while a ransom was worked out with the shipping company. This would then be paid in cash and delivered in a parachute drop to the ship or nearby water. Approximately 3 million dollars each was paid for the release of M/V *Faina*, *Sirius Star* and the Danish ship CEC *Future* and their crews in 2008–9.

Fatigue can be an issue, especially when the ship is vulnerable just after leaving port when the sleep rhythms of the crew have been disrupted by work and entertainments on shore. This is not a new phenomenon, but it rises in importance when the crew numbers are reduced. For even the very large ships it is common to sail with as few as 17 crew instead of the once typical 44.[35] The IMO recommends turning the ship into a citadel when in pirate waters. The crew are positioned inside and all doors are securely locked. This is inconsistent, of course, with the desire to have crew watching on the decks to deploy measures to keep pirates from boarding. In addition as one ship's captain said: 'If someone throws a firebomb on deck and you are locked inside, then what do you do? You are trapped and the ship is on fire. At least some of my crew were able to put out the fires while the pirates were aboard. I think the citadel is a stupid idea.'[36]

What is happening ashore influences piracy. The disintegration of government in Somalia is playing a significant role in the spike in piracy there. Illegal fishing and dumping of waste has also contributed to destroying local fishing opportunities. Nets from foreign trawlers destroyed corals that were the habitat for lobsters traded with Dubai for big profits. Somali pirates initially gained credibility and even heroic status with the local population by presenting themselves as coastguards.[37] Extreme poverty can also feed desperation. Somali piracy fell away for a month from mid-December 2008. Without mentioning the international response, which I will return to later, the IMB suggest that the resumption of piracy on January 26, 2009 was due to 'favourable weather conditions in the area and the high number of hijacked vessels that have been released recently'.[38]

The organisation of pirate activities into syndicates is a new phenomenon. In Somalia until June 2006, the waters were divided into sections under the control of a syndicate leader with gangs of up to eight working at sea, mainly in an attempt to get ransom money.[39] After a lull in 2007, pirates became well organised again. By 2009 there were four main groups, one including former navy sailors who used a naval patrol boat as their mother ship. Links with onshore organisers have also strengthened.[40]

Burnett claims that Asian pirates often now link in with four international crime organisations: 'The Singapore syndicate controls the southern part of the South China Sea and Malacca Straits.

Bangkok controls the Andaman Sea bordered by Thailand, Burma and Malaysia; Hong Kong controls the northern parts of the south China Sea; Jakarta controls the Java Sea and parts of the south China Sea to Borneo.'[41] These syndicates are in a better position to make complex ransom demands than a small band of poor attackers. They are also well placed to organise the disappearance of one ship and the appearance of another (painting over all distinguishing marks; re-flagging the ship, producing new sailing papers and substitute crew). This greater organisation leads to increasing profits and more funding for future piracy.

THE RISE OF PIRACY IN SOMALIA

Somalia experienced a rise in pirate attacks in 2005 with 35 recorded incidents. A remarkable range of vessels was targeted. In June 2005 the *Semlow* was loaded with 850 tons of rice and food aid, chartered by the United Nations to help the Somalis hit by the Indian Ocean tsunami. Fifty-five kilometres off Somalia, the *Semlow* was boarded by heavily armed pirates. They were made aware of the nature of the humanitarian mission, but this did not deter them in their quest to get $500,000 in ransom for the ship and crew. The pirates took the *Semlow* close to shore off the central Somali town of Ceel Huur. They retained control of the ship for four months eating the World Food Program rice and other fresh food bought to them by collaborators on shore. The crew had survival rations of food and water. During these months some of the pirates took off in speedboats and hijacked the *Ibn Batuta*, an Egyptian ship carrying cement. The two ships then sailed towards Harardhere. The captain of the *Semlow* was taken ashore to meet the pirate bosses and given letters from his wife. He was returned to the *Semlow* and the pirates left a few days later, after a $135,000 ransom was paid. The Somali negotiators sent out a boat to escort the *Semlow* into port and the rice was off loaded. On returning home the captain radioed the *Torgelow*, a sister ship, and discovered that he was talking not to the captain but to the pirate chief who had taken over the *Semlow*.[42]

In November 2005, a passenger ship, the *Seabourn Spirit*, was attacked while passing the Somali coast. The pirates fired machine guns and a rocket-propelled grenade, which failed to explode. They didn't succeed in boarding the boat due to clever manoeuvring by the captain and the use of a long-range acoustic device that blasts ear-splitting noise in a directed beam. It was at this point that the

IMB said Somali waters were 'out of control', but there was a great deal more to come.

In March 2006, 25 nautical miles off the Somali coast, a US guided missile cruiser and a guided missile destroyer, part of the US navy's Fifth Fleet based in Bahrain, were conducting maritime security operations as part of a Dutch-led task force. These operations were not related to piracy. However, when the Americans saw a suspicious looking 30-foot fishing boat towing two others, they sent out zodiacs to investigate. The zodiacs were fired upon by the suspects. Gunners on the destroyer then lay down covering fire to allow the naval crews to escape. One suspect was killed and five wounded. Those still alive were taken into custody. Defending their actions the Americans said they were in international waters which are 'under international law'. This was disputed by the Somali Justice Advocacy Centre. A spokesman for the Fifth Fleet, Cmdr Jeff Breslau, said: 'Why the men opened fire is unclear, but their decision to take on navy ships in a 30-foot fishing boat was not too smart ... If somebody shoots at us, they can pretty much expect to die because we will return fire.'[43]

During 2006, pirate attacks in Somalia stopped. This was attributed to the rising influence of Islamic rule and shariah law briefly ending the control, or lack of it, exercised by competing warlords. The punishment for piracy under shariah law is execution or amputation. Late in 2006, the Islamic government was disbanded due to the intervention of Ethiopian forces. Mogadishu fell into chaos as rival clan militiamen seized back power. A UN-backed interim government remained weak, itself torn apart by clan rivalries. Early in 2007 piracy was on the rise again with Somali attacks on a diverse range of vessels off the east coast and in the Gulf of Aden, including tankers, dhows, cargo vessels, bulk carriers, containers, general cargo ships, yachts, research vessels and tugs.[44] By 2007 the number of reported attacks off Somalia had risen again to 31 together with 13 in the Gulf of Aden, almost to the 2005 level. Piracy operations became more efficient with the use of fast skiffs launched from mother ships. The skiffs were blue or white to blend in with the ocean vista. Sophisticated weaponry was used.[45] By the end of 2008, piracy attacks in Somalia were almost a daily occurrence, sometimes with multiple attacks on one day, for example, October 28. The figures for the first three quarters of 2009 for actual and attempted attacks exceed the total for 2008 and the weather beyond that date became more favourable to piracy. A wide range of vessels continued to be targeted sometimes at great distances from the shore, well into the high seas. In late 2009 there were several attempts by Somalis to take over ships approximately 1000 nautical miles from the Somali coast. Some ships were damaged by rocket-propelled grenades. Two hijacks occurred in mid-November: a bulk carrier and a tanker with 22 crew members taken hostage in one case and 28 in the second. The captain of the tanker died of wounds sustained in the attack.[46]

After many hijackings several countries sent naval forces to the region including Russia, the US, Denmark, France, Greece, Germany, Spain, the Netherlands, the UK, Iran, Canada, Pakistan, Malaysia, India and China as part of or in addition to NATO and EU naval vessels or air forces. (Australia did not send forces but the frigate HMAS *Sydney*, when passing through the Gulf of Aden on another mission, rescued the merchant vessel *Dubai Princess* from a pirate raid.) Permission was sought from the UN to move into Somali territorial waters. In June 2008 the Security Council passed resolution 1816 authorising states co-operating with Somalia's transitional government to enter its territorial waters.[47] This resolution had the agreement of the Somali Government yet months passed and no foreign states acted on it. More and bigger ships were hijacked. Huge ransoms were paid. Shipping in the major routes connecting Europe and the Middle East and Asia was disrupted. A NATO naval blockade guarding a narrow shipping lane for 600 miles passing the Somali coast had some success in curbing piracy. Planes and helicopters tracked suspicious vessels that were then boarded and searched. In December 2008 the EU sent naval frigates and aircraft, with the aim of implementing the UN resolution. Before their arrival, in November 2008, the UN adopted another resolution to try to combat Somali piracy. This called for travel restrictions and an asset freeze to be placed on people and organisations threatening Somalia's peace and political process, or obstructing humanitarian assistance. This was heralded as taking a 'hard stance',[48] but in a lawless country what impact could it have?

By early 2009 the EU forces had set up a longer guarding corridor than the NATO forces. There was a brief lull during this time of intense international activity in the Gulf of Aden and the Indian Ocean off Somalia, though pirates still held 15 ships and over 200 crew members. The 'armada' had succeeded. Or had it? From January 26 the attacks resumed. Was the quiet time just due to the weather? How could such a huge effort fail? One of the problems in addressing Somali piracy is that the counter forces do not always act in an international spirit. Some naval ships are not giving assistance to vessels flying particular flags or carrying crew from certain countries.[49] This points to a need for a central authority to co-ordinate piracy patrols. Such an authority could also facilitate speedier responses and introduce long-term strategies more effective than the short-term plans currently in place.

A second problem is that there is no clear policy for dealing with captured pirates. This relates back to the jurisdictional issues discussed above. The Law of the Sea states that 'The courts of the State which carried out the seizure may decide on the penalties to be imposed.' However, captured Somali pirates are usually returned to Puntland in northern Somalia, either from the belief that they are the responsibility of the Somali Government or an unwillingness to transport the pirates to the home country for prosecution. Some states are concerned that pirates may seek asylum if they are taken to a foreign country for prosecution; or the state that captured the pirates may have no laws under which to try them. Some states are worried about returning pirates to Somalia if shariah law is strictly followed and the pirates face death sentences. By early 2010 this law is supposed to be in place but is not enforced. The returned pirates are generally not detained.

These are some of the fears that inhibit action, and when counter-forces do act, apart from just taking the pirates home, there is inconsistency. The French transported some pirates to France where they remain in custody. (The pirates argued that their transfer to France was illegal.) Some other pirates captured by the French have been sent to Kenya for trial. Kenya has agreed to take a limited number of the pirates captured by the EU, US and UK forces.[50] The Spanish handed over nine pirates to the authorities in the Seychelles. The US took one of the pirates who held Captain Phillips of the US-flagged *Maersk Alabama* hostage back to the US for trial. The legal confusion over prosecution is given another twist in the comments by Lieutenant Commander Alexandre Fernandes from the NATO anti-piracy patrol. Discussing the incident where Dutch commandos overpowered pirates and took control of their mother ship he says:

> We have freed the dhow [mother ship] and we have seized the weapons ... The commandos briefly detained and questioned the seven gunmen, but had no legal right to arrest them ... They can only arrest them if the pirates are from the Netherlands, the victims are from the Netherlands, or, they are in Netherlands waters.[51]

Likewise, the Portuguese navy disarmed pirates who attempted to board the Greek tanker *MV Kitton* in May 2009, but then released them reasoning that they could only make arrests if the attack was against Portuguese crew or ships.[52]

Clearly there is a major problem of ocean governance surfacing here. A central authority with the power to oversee prosecutions is needed. This should be put in place with some urgency, not only to help decrease the number of pirate attacks but also to ensure fairness. In 2008–9 an increase in violent responses to piracy occurred. To mention just three examples: (1) three pirates holding Captain Phillips in a small boat were shot dead by US forces; (2) French commandos killed two pirates and one hostage when seizing control of a pirated yacht; (3) the Indian navy destroyed a pirated Thai fishing trawler believing it to be a pirate mother ship – 15 Thai fishermen died in the attack. How can these actions be justified? A UN Charter signed after the Second World War aimed to banish the use of force or the mere threat of force in international relations except for self-defence. This is arguably a justification in example (1), though the captain had a high value for the pirates alive and no value dead. Self-defence could also be used to justify (2), but is it better to risk the loss of innocent lives than to pay a ransom? There is no possibility of giving the pirated crew any decision-making powers in such a situation. The least justifiable action is (3). However a naval spokesman said that the Indian warship was acting in self-defence, as armed pirates were visible on the deck.[53] The comment is incredible given that the naval vessel was 400 feet long and fully equipped for battle. Clearly there needs to be more thought given to this escalation of violence and what the limits should be. Also the tragedy of the Thai trawler could have been averted. The IMB had notified forces in the area that the Thai boat had been captured but the Indian navy has no direct communication links with the IMB – a situation graphically revealing the need for central governance.

The signs of a tough justice are emerging in Somalia too. Vigilante groups are now going after pirates who are losing their heroic status.[54] They have been handed over to the local authorities. If the law is enforced they will face the death penalty.

In 2009 a detectable shift occurred. In 2008 the actual piracy attacks off Somalia easily outnumbered the attempted attacks. In the early stages of 2009 these figures were reversing, and this continued throughout the year.[55] However, addressing Somali piracy is not just about surveillance and prosecution. While the state is in such disarray, while hunger is widespread and employment possibilities are very limited, the attractions of piracy will remain. An international authority would be well positioned to work out some positive initiatives to deal with captured pirates, rather than incarceration or death, initiatives that might go some way to alleviating the internal problems of countries that are unable or unwilling to address piracy.

PIRATE ATTACKS ON PRIVATE BOATS

The two main piracy reporting agencies mentioned above, the IMO and IMB, do provide statistics for attacks on private boats, but there appears to be extreme under-reporting. Since 2004 the IMO has included attacks on private boats in its piracy statistics but not many are listed. The IMB provides yearly reports on yacht and speedboat attacks and lists 105 incidents from 1993 to September 2009.[56] An organisation called the Information Centre for Bluewater Sailors collected reports on piracy attacks on yachts from 1996 to October 2009 and documents 148 incidents during that time, mainly in waters off Yemen, Venezuela, Brazil and the Caribbean. The Centre suggests that even this is likely to be less than the number of actual attacks.[57] An internet site, *The Global Site for Cruising Sailors*, is a good source for first-hand reports on piracy and the organisation of convoys for protection.[58] In 2008 this site began listing attacks on yachts in a 'piracy round-up for the year'. The *Global Site* was notified about 44 attacks in 2008. Reports from Venezuela, the Caribbean, Gulf of Aden/Somalia and the Solomon Islands were the most common. In contrast, the IMB listed only nine attacks on yachts in 2008. The IMB should more accurately report piracy against yachts given that it is the main documenting agency. Its gross underestimate is very misleading. It seems that very few want to take yacht piracy seriously. Until very recently there has been a lack of coverage of the issue in sailing magazines, signalling that piracy 'is a kind of taboo which apparently doesn't fit into the image of an intact world of bluewater sailors'.[59]

As in the case of commercial shipping, pirate raids on yachts can be focused in one area and then quickly cease only to surface somewhere else. Papua New Guinea was listed as safe in 2003 by the Information Centre for Bluewater Sailors. After this time attacks increased, often involving violence with machetes and guns.[60] The Information Centre notes that ten years ago the Venezuelan coast had no pirates but that since then attacks have occurred quite frequently. A Swedish man was shot in the stomach and a woman threatened on their Swedish cruiser *Miren s/v Lorna* in 2001. After putting out a distress call, the woman, Lorna Vivi-Maj, refused the aid of the Venezuelan Coast Guard. This Coast Guard had previously tried to raid a German yacht. In her report Lorna does not say she was aware of this, but she was clearly suspicious. She accepted help from the Trinidad and Tobago Coast Guard and her partner survived.[61]

In 2004 the Venezuelan Coast Guard did act to find a yacht in their waters, the French yacht *Les Chouans*, after the captain had failed to make contact for several days. He was sailing alone. The Coast Guard located the yacht floating in a cave, finding the captain dead from a gunshot wound to the head. However the crew of another yacht attacked and robbed in Venezuelan coastal waters in 2005 received no official support for eight hours, despite being at anchor only 'a hundred yards or so from the Carenero Yacht Club'.[62] In 2008 there were more reports of attacks on yachts in Venezuela than anywhere else. In once incident the British-flagged yacht *Raven Eye* was boarded by armed pirates. The crew were injured and the skipper's dog was shot and stabbed while acting in their defence.[63]

The sea territory around Somalia and Yemen is dangerous for yachts as well as commercial vessels. The audacity of the armed hijack of the luxury yacht *Le Ponant* in 2008 grabbed world headlines. This occurred in the Gulf of Aden and the boat was taken to Somali waters. The boat and crew were released after a few days and a ransom payment. Also in 2008 a French-flagged yacht, *Carre D'AS IV*, was similarly hijacked. French commandos captured the yacht, killing one pirate and detaining six others.[64] In a tragic encounter in 2009 one of the crew of the yacht *Tanit* was shot dead, along with two Somali pirates, by French forces in their effort to rescue the hostages.[65] An English couple had their yacht hijacked by Somali pirates near the Seychelles in October 2009. The couple were taken to Somalia and a ransom demand for $7 million was made. Late in 2009 negotiations continued with the possibility that the couple may be exchanged for seven captured pirates instead of the ransom.[66]

TERRORISM ON THE SEA

Piracy, by definition, is for private gain. The hijacking of a ship on the high seas or attacks on ships' crew or passengers in these waters, or the stealing of goods from ships on the high seas, if done for political reasons, is not called piracy. Seeking ransom money for ships or crew also isn't piracy if the act is committed for political ends. This sounds simple but it is not always easy to answer the question what counts as a political end. If fishermen steal from boats to feed their community is this political or private? It seems to be political but then political attacks on ships are often labelled 'terrorist', and it would seem absurd to label small-scale political attacks in this way.

Antonio Cassese attempts to capture the legal definition of terrorism as follows:

> Any *violent* act against *innocent people* intended to force a state, or any other international subject, to pursue a line of conduct it would not otherwise follow is an act of terrorism. Such acts are prohibited both in times of *peace* and in cases of *armed conflict* whether civil strife, a war of national liberation or an armed conflict between states.[67]

This definition allows us to say that the fisherman's actions were political but not terrorist. It leaves them in limbo as they are not acts of piracy, lacking the private gain. This issue will be left unresolved. It highlights just another of the difficulties with the Law of the Sea definition of piracy. The Cassese definition has a weakness in that violent offenders attempting to get *a state* to pay ransom money for the hijacking of a ship, say, would count as terrorists, even though their actions might be for their own personal gain. To further highlight the complexity of terrorism at sea, I will outline an apparently clear case and then reveal the confusion in practical and legal responses.

Antonio Cassese's study of the *Achille Lauro* hijacking by terrorists shows how Italy, Egypt, the US and the PLO worked sometimes in unison, sometimes not, in order to minimise risk to captured civilians and to bring the hijackers to court. Cassese argues that the outcome was successful despite one murder, some of the ringleaders escaping and violations of international law.[68]

On October 7, 1985, four members of the Palestine Liberation Front, a faction of the PLO, took command of the *Achille Lauro*, an Italian cruise liner, close to the Egyptian coast. There were fewer passengers on board than normal, only 201, as 600 had left the ship at Alexandria to visit the pyramids with the aim of re-joining the ship in Port Said. There were 344 crew.

The hijackers were heavily armed. After taking over the boat they threatened to kill the passengers unless the Israelis released 50 Palestinian prisoners. The passengers were herded into the main saloon. Several cans of kerosene were placed there with the threat that they would be lit if the hostages caused trouble. The hijackers forced the crew to head for Tartus in Syria. Twenty Americans and British nationals were separated from the other passengers and crew to form the first targets. There were also Swiss, Austrians, Italians and Germans on board.

Concerned states including Italy, Egypt and the US had crisis talks. The Italians led the discussions and negotiations, as the *Achille Lauro* was an Italian ship. However, the Egyptian authorities were also deeply involved because the ship was hijacked in Egyptian waters. The US was concerned because of the number of Americans on board. The PLO at the outset dissociated itself from the hijacking, 'condemning it as an act of sabotage against its own peace efforts',[69] and offered to help in the apprehension of the hijackers. The PLO Penal Code provides for the imprisonment of Palestinians who carry out hijackings of vessels belonging to friendly or foreign states.

On arrival at Tartus, the Syrians were not prepared to negotiate with the hijackers. All the other ships left the port. The captain of the *Achille Lauro*, Gerardo De Rosa, said later that this was 'something unexpected and sinister', thinking that the Syrian authorities had told these ships to leave. 'In no time at all we were alone: the sea around us was completely empty.'[70] Unquestionably the captain feared a military strike against the hijackers.

On October 8, one American passenger was murdered by the hijackers, who then ordered the captain to make this public to try to force negotiations. They sailed for Libya. The US were ready to begin military action on the night of October 8/9 but the Italians wanted to pursue peaceful solutions first. They stressed that as the *Achille Lauro* was Italian, only Italy would have the right in international law to take military action, if the need arose. Also on October 8, Abul Abbas entered the negotiations, ostensibly as part of the PLO. He travelled to Egypt, and the hijackers, pleased with this development, changed the course of the *Achille Lauro* back to Port Said, anchoring 15 miles off the coast on October 9. The PLO tried unsuccessfully to get Italy to urge Israel to release at least some prisoners. Then the PLO leader Arafat asked the Egyptian authorities to transfer the hijackers safely to the PLO headquarters in Tunis for trial with 'all passengers safe and well'.[71] It seems that the one murder that had occurred was held out of view. The

Egyptians, Italians and Germans initialled an agreement to give the hijackers safe conduct on October 9. The Americans and English refused to do so. Cassese notes

> By 3.30 p.m. that day, the hijacking was all over and an Egyptian tug moved alongside the liner to take off the four hijackers. The nightmare seemed to be over. From on board the tug, an Arab (in fact Abul Abbas) waved to the passengers and said how sorry he was that the last three days they had been put to such inconvenience.[72]

A few hours later the murder was public knowledge. It is possible that the captain held back this information to get the hijackers off the ship with a negotiated plan and to assure the safety of the other hundreds of people on board.

With this 'new' information, Italy wanted to put the hijackers on trial for murder but Egypt allowed them, along with Abul Abbas, another PLO representative and ten armed guards, to board a plane bound for Tunis. The pilot was refused permission to enter Tunisian air space. The Americans had exerted pressure on Tunisia for this outcome. The plane was flying back to Cairo when, under orders from President Reagan, it was intercepted by four fighter planes and forced to land at a NATO base in Sigonella in Sicily. The fighter planes didn't land but two US military transport planes did. As soon as the Egyptian aircraft landed it was surrounded by 50 Italian soldiers from the base who were surrounded by 50 American soldiers from the transport planes. They were instructed from the White House to arrest the hijackers and take them to the US. Bettino Craxi, the Italian prime minister, argued that 'the crimes had been committed in international waters, on board an Italian vessel, and should therefore be considered criminal acts perpetrated on Italian territory'.[73] President Reagan was forced to give in.

The Egyptians agreed that the four hijackers could leave the plane and be put under arrest but they did not force the two PLO representatives to leave the plane. By now there were suspicions about these representatives, so they were to be taken to Rome for further questioning. The plane took off tailed by an American plane and four Italian fighter jets. Cassese conjectures that these escorts were to reassure the Americans that the plane wouldn't return to Egypt and to reassure Egypt that the Americans wouldn't intercept the plane again. The legal authorities in Rome could not find any grounds to detain the PLO officials. They were sent home on a Yugoslavian airline.

The four hijackers all received jail sentences from the Italian courts, and when more information came to hand Abul Abbas received a life sentence in absentia for his role in the hijacking. Another ten people were also tried and found guilty over the incident. The PLO and Arafat were cleared of any involvement.

During the complex web of negotiations, plane flights and forced ship excursions, some principles of international law were violated. Egypt, knowing that a murder had taken place still allowed the hijackers to fly out of the country. Yet Egypt is a party to the 1979 Convention on the Taking of Hostages. This Convention required Egypt to put the hijackers on trial or to extradite them to the US given that the murdered man and numerous passengers were Americans. Either move went against political expediency.

In an attempt to force a desired destination on the hijackers once they were in the Egyptian plane, the Americans illegally intercepted the plane in international air space and later went on to violate Italian air space. These actions go against the charter on the use of force in international relations as mentioned above. No appeal to self-defence could be made once the hijackers were off the boat and the hostages freed. Yet this was when the Americans used force. Also by now the Italians had already expressed a desire to try the hijackers and would have sought extradition from Egypt or Tunisia had they landed there. So the use of force was unwarranted practically as well as legally.

Cassese argues that the Italian Government also violated international law in this case. They did not abide by the 1983 extradition treaty. The Americans believed they had sufficient evidence against Abul Abbas to show that he had masterminded the whole operation. The extradition treaty lays down that when one of the parties requests an extradition for crimes not committed on its territory then the other party has discretion whether or not to grant the request. The Italian judges decided that the evidence against Abbas was insufficient to arrest him. The Americans argued that they were not given a fair opportunity to present this evidence. Convincing documents arrived the next day but Abbas was already free. Delaying a decision would have been more consistent 'with the *spirit* of the treaty with the United States, and the *principles of friendly co-operation* that should exist between states, especially when they are allies'.[74]

The Italian court was working with an idea of terrorism as a political act involving systematic resort to particularly violent means such as to generate panic in the general public. The judges decided that the hijackers were terrorists. Cassese agreed, but argued that the various state responses to this terrorist attack highlight the weakness of international legal regulation concerning how states should respond to terrorism on the sea.

As a result of the *Achille Lauro* incident, a new Convention was drafted by the IMO called the Convention for the Suppression of Unlawful Acts Against the Safety of Maritime Navigation. The aim of this Convention was to assist the fight against piracy and terrorism at sea. The unlawful acts include 'seizure of ships by force; acts of violence against persons on board ships; and the placing of devices on board a ship that are likely to destroy or damage it'.[75] No distinction is made between private and political motives. Piracy and terrorism on the sea are treated the same. The location of the offences isn't specified, so there is no necessity that they occur on the high seas.

The Convention obliges contracting governments to put in place preventative measures and to either extradite or prosecute alleged offenders. The Convention came into force in 1992, and a Protocol was added in 2005. This covers offences against ships or passengers and crew involving explosive, radioactive material or biological, chemical or nuclear weapons when the purpose of the act is to 'compel a Government or an international organisation to do or to abstain from any act'. It is also unlawful to intentionally discharge hazardous or noxious substances that are likely to cause death or serious injury or damage. It is an offence to use a ship in a manner that causes death or serious injury or damage.[76] This is an attempt to cover terrorist acts on the sea and is phrased so as to ignore small-scale political piracy. The 2005 Protocol has not yet been ratified.

There are problems in the application of the Convention. Nothing is said about whether offenders can be pursued into territorial seas. A coastal state may not have the means or inclination for pursuit and may refuse permission for foreign law enforcement ships to enter their waters. If these problems can be overcome or if a ship is out of territorial waters then representatives of a state signed up to the Convention may board a ship if there are reasonable grounds for suspicion that an offence under the Convention has been committed. However the authorisation and co-operation of the flag state is required before a boarding can take place. Ships flying a 'flag of convenience' may be immune from inspection because the flag state may be unwilling to co-operate. In 2009 some states commonly used as flag states had not signed the Convention, including the Dominican Republic and Belize. In any case the Protocol is (as of 2009) not yet in force.

The Law of the Sea is silent on terrorism. Yet the *Achille Lauro* case does not stand alone. A US naval research vessel, the USS *Liberty*, was attacked by Israeli forces in 1967 killing 34 crew and injuring 170 more.[77] The description of this incident as a terrorist attack is, however, disputed. In 2000 the naval destroyer USS *Cole* was damaged by suicide bombers. Seventeen crew were killed and 39 others were injured; Al Qaeda took responsibility for the attack.[78] M. C. W. Chong points out some recent terrorist activities in Asian waters. The ferry *Our Lady Mediatrix* was sunk in 2000, allegedly by Moro Islamic Liberation Front insurgents; kidnappings occurred from the tanker *Tirta Niaga* in 2001, by the Free Aceh movement who were also responsible for kidnappings from the *Ocean Silver* tug in the same year. Suicide bombers linked to *Jemaah Islamiah* came close to blowing up a US warship in Singapore in 2001.[79]

The Convention attempts to cover piracy and sea terrorism though the Protocol is more explicit on the latter. These instruments are not, however, acting as effective deterrents. In 2009 Somalia was not a signatory to the Convention, nor were Malaysia and Spain who sent in combat forces. The Law of the Sea and the Convention/Protocol seem to have little relevance in this area. However, certain raids on the sea have given an impetus to efforts to tackle piracy. The hijacking of the *Alondra Rainbow* led to the first prosecution for piracy under the Law of the Sea. The hijacking of the *Achille Lauro* led to an attempt to tighten the legal regime for dealing with pirates and terrorists on the sea by setting up the Convention and Protocol. The spike in piracy attacks in the Straits of Malacca led to quite successful regional agreements. The current Somali situation is likely to prompt action also. The UN resolution allowing foreign vessels entry into Somali waters has been extended. However, most of the problems are occurring much further out in the Somali EEZ or even the high seas. Other initiatives need to be investigated but new measures should not just be put in place for one country leaving the possibility of having to begin all over again if major piracy incidents arise in other places. The reflections in this chapter lead to the view that waters beyond the 12 nautical mile limit should be given over to an international body for management of piracy patrols and co-ordination of international forces. Such a body should also take responsibility for prosecution and rehabilitation through a central court to try to achieve uniformity and fairness.

While conflicts between pirates and those who pursue them are currently in the media spotlight, some other battles on the seas have not received such attention even though they pose a far greater environmental and consequently human risk. These concern the over-exploitation of fish.

5
The Fishing Wars

I speak for those who have no voice – the fish.

Brian Tobin, aka The Tobinator,
Canadian Minister for Fisheries and Oceans, 1995.

The case for a change in ocean governance is building. To fairly and responsibly manage undersea non-living resources and heritage an international body is needed. To have a chance of addressing the problems of piracy on the scale now seen we need an international body with unimpeded rights of access to all waters beyond the territorial seas and powers to co-ordinate patrols and prosecutions. At present, coastal states have sovereign rights to fish out to 200 nautical miles. These rights are incompatible with the proposals suggested as the waters beyond 12 nautical miles would be owned by all and governed in the interests of all. They would be international in a fully fledged sense. Are there good reasons to maintain the sovereign rights to fish that presently exist? A positive answer seems to emerge from an analysis of the recent fishing wars.

THE COD WARS

The Dutch, French and British fished extensively off Iceland from the eighteenth century on but the threat to the cod population escalated with the arrival of the British steam-powered steel trawlers. These vessels were hugely successful in destroying fish populations in the North Sea in the early twentieth century within ten years. Mark Kurlansky, in a remarkable book entitled *Cod: A Biography of the Fish that Changed the World*,[1] notes that these trawlers ran over the nets and lines of Icelandic fishers still using unsophisticated vessels. The Icelandic response was to acquire these new types of vessel. Germany began fishing off Iceland in the 1920s, even within three miles from shore, using coded messages between boats to evade the Coast Guard. If it had not been for the halt to fishing during the two World Wars, Icelandic cod may have been fished out of existence.

In 1952 Iceland began extending its sea territory, claiming exclusive fishing rights out to four nautical miles, an extension of one nautical mile from the agreed upon limit at the time. In 1958 Iceland extended the border to 12 nautical miles. This led to confrontations with the British trawlers who continued to fish within this zone. One clash between an Icelandic naval ship and a British trawler was thwarted by a ship from the British navy called the HMS *Russell*. The captain of HMS *Russell* also threatened to sink an Icelandic boat which was ready to fire on a British trawler.[2]

An uneasy peace was secured when both sides agreed to accept decisions from the International Court of Justice on fishing grounds. A new government in Iceland in the 1970s rejected this agreement and expanded Iceland's exclusive fishing zone to 50 nautical miles causing further conflicts with British trawlers and skirmishes between the Icelandic Coasts Guard and the British navy. By modifying minesweeping technology the Coast Guard developed a device to trap trawl cable and cut it, releasing the fish in the trawl. This was very effective. The British and German boats would attempt to evade the trawl cutters by ramming the Coast Guard boats but they were stronger, built as icebreakers, so could always win these battles. Hannibal Valdimarsson, the Icelandic Minister of Communications, called for port authorities to refuse aid to British ships in Icelandic ports. The Government of Iceland blocked NATO planes from Icelandic air traffic control. There were threats to cut diplomatic relations with Britain. In 1974 the International Court of Justice ruled that it was not permissible to hinder foreign fleets in the contested area off Iceland. Iceland ignored this ruling.[3]

Then in 1975, with herring populations crashing and cod populations reducing by one third and facing 'imminent ruin',[4] Iceland declared a 200 mile exclusive fishing zone. This move was a further violation of international law that was eventually modified to fall in line with the decision in 1994. In the 1970s the British did not recognise Iceland's authority to extend its fishing zones and continued fishing within the 200 nautical miles, assisted by the British navy. The navy tightly patrolled around the edge of the fishing fleets keeping off the Icelandic Coast Guard who attempted, sometimes successfully, to cut the nets of the British trawlers. Ramming occurred on both sides and shots were fired causing some injuries to crew and boat damage. Expressing the idea that this was a restrained military conflict, the British foreign secretary said: 'Both sides in the conflict are showing valour, but there is no need for anyone to show their virility.'[5]

Fearful that they would loose their last viable fishing population, Iceland refused to listen to the International Court and broke off diplomatic contact with Britain. Iceland had one drawcard, a NATO base at Keflavik that was strategically very important in the defence of the Atlantic Ocean from the Soviet Union. The Icelandic Government threatened to close this base if the foreign fishers did not leave the 200 mile zone. Britain bowed to this pressure and withdrew.[6] In 1976 the European Economic Community declared a 200 mile exclusive fishing zone, undercutting British or German claims concerning a right to fish within 200 miles off Iceland.

Iceland's willingness to engage in extreme action to protect its fish populations now takes the form of very stringent internal conservation controls. Glover, in the book *The End of the Line: How Overfishing is Changing the World and What we Eat*,[7] ranks Iceland as managing the second most successful fishery in the contemporary world, after the Atlantic salmon fishery. The latter involved buyouts of salmon fishing interests in Greenland, the Faroes and North East England and has allowed salmon to return to rivers in Iceland and the River Tweed on the English/Scottish border.

Property issues come up in the management of the Icelandic cod. Areas where juveniles may be caught are closed off and this is written into the law. There are strict limits put on the total allowable catch too. The key to the recovery of the cod populations resides in a rights-based system of quotas. The quotas give fishers 'a right to take a certain proportion of the stock, depending on scientific advice, for the foreseeable future'.[8] Fish then come to be viewed as the property of the fishers who have a stake in the health of the fish populations. This had led to a changed attitude to fishing where conservation is in the forefront. The way people fish has changed too. Without the drive to catch the most fish, the effort goes into catching high quality fish and selling them for the best prices. The rules are strictly enforced by government agencies and the Coast Guard with heavy penalties for non-compliance including withdrawal of fishing licences and jail terms. However there is some flexibility. If a fisher has caught more fish than their quota, they may buy extra quota. If that fails then the fish are landed but 80 per cent of the value of the over-quota catch goes to the Marine Research Institute.[9] Ownership then is one way to give value to fish and to motivate conservation. Iceland's strong stand has saved the cod in its waters.

By contrast, the cod off the coast of Newfoundland were fished too hard for too long and their recovery is not predicted. In 1497

Giovanni Caboto sailed east from England looking for Asia. He stumbled on Newfoundland. His voyage was for Henry VII of England and the English knew him as John Cabot. He remarked that the Newfoundland waters were teeming with cod, so easy to catch that one only had to lower a basket with a stone into the water.[10] The French, English and Spanish fished these waters from the sixteenth century. The cod were salted or dried for export. Kurlansky points out that dried cod was called 'Poor John' which was also a name Shakespeare used in *The Tempest*: 'He smells like a fish – a very ancient and fish-like smell – a kind of, not of the newest, poor John'.[11]

The abundance of cod in these waters was maintained by vast quantities of phytoplankton, zooplankton and krill. The cod hatchlings fed first on the phytoplankton then on the zooplankton followed by the krill. Then they headed for the bottom to hide from predators from whom they would not be safe for about a year. They were caught in bottom draggers, huge nets trawling over the ocean floor, a common technique used widely from the nineteenth century on.[12]

The number of cod taken greatly increased in the 1960s with more sophisticated foreign vessels. The populations declined from 1968 and in 1977 Canada declared a 200 mile exclusive fisheries zone. Although foreign fishing decreased, the Canadians took over fishing in this zone with powerful, offshore boats and misguided scientific management.[13] Also cod migrate in and out of the 200 mile zone so foreign fishers could quite legally stand off in the high seas beyond 200 nautical miles and fish the cod. This they did until the population dropped below sustainable limits. A moratorium was imposed in 1992, lifted in 1995 in an attempt to solve the social problems arising from unemployment, but the fish population had not recovered and the fishery was closed again in 2003.

The possibility of biological extinction of the Canadian cod is real. It is now clear that with fish, depleted populations become less productive. Also cod tend to huddle together when their numbers are small. In the winter of 2003 cod shoaled in Smith Sound, Newfoundland. About half the shoal was pushed into the coldest water where ice was forming. They became smothered when ice blocked their gills and they weren't able to extract oxygen from the water.[14]

The Canadian cod problem was thought to be generated to a significant degree because the Canadians had no ownership rights over the seas which the foreign fishers were exploiting and they were unwilling, unlike Iceland, to make a move to expand sea territory until 1977, when the cod population collapse was already well under way. Their management strategies in regards to quotas were in retrospect scientifically flawed and were undermined anyway when the cod swam into the high seas. International agreements to limit catches in these seas did exist but they were disregarded and unenforceable. Crucially they could only be enforced by the flag state of the vessel illegally fishing and that flag state, where the vessel was registered, might have been thousands of miles away. In addition, the flag state is likely to have been 'convenient' in relation to illegal activities.

THE TURBOT WAR

In 1986 Spain and Portugal joined the European Community. One condition of membership was that their fishing fleets move away from the coasts of other European members for ten years to save traditional fishing grounds from over exploitation. Portugal then took up fishing off Newfoundland and Spain increased its presence there. After the cod collapse the Spanish and Portuguese fleets targeted turbot, also called Greenland halibut, which led quite quickly to a military confrontation with Canada.

Brian Tobin, the Canadian Minister for Fisheries and the Oceans in the 1990s, waged an incredible rhetorical campaign to save the turbot, including those beyond Canada's 200 nautical mile zone.[15] The EU fishers rapidly depleted turbot populations which swam from Canada's EEZ into the high seas on migration routes. The Northwest Atlantic Fisheries Organisation (NAFO) is a UN body that is supposed to look after conservation in the high seas off Canada. Their inaction aided the demise of the cod. A couple of years after the 1992 moratorium on cod, NAFO warned that turbot were being fished too intensively but they didn't do anything quickly to stop this practice. Canada reduced its own quotas in the 200 mile zone and urged NAFO to do the same in the high seas. No action was taken until 1995.

Tobin argued that Canadians had held down their catches for several years for reasons of conservation 'while the EU has taken too much'.[16] He strongly criticised the Spanish and Portuguese fisheries for not controlling illegal fishing in their fleets. He warned the EU ministers that 'if the EU could not prevent its fishing fleets from breaking NAFO regulations, Canada was prepared to do so even on the high seas'.[17]

In 1995 quotas for the EU fleets going after turbot were severely cut. This angered Spain and Portugal. Tobin responded: 'There is no reward for those who ignore the needs of conservation',[18] and warned of serious consequences if they didn't follow the NAFO decision. The Spanish fisheries minister objected to Canada's attempt to influence fishing on the high seas and spoke about a rupture in transatlantic relations. The EU fisheries commissioner, Emma Bonino, insisted 'We are not the pirates of the Atlantic',[19] but she objected to the quota allocation. The EU voted to intervene to help lift the quotas for Spain whereupon a Canadian fisheries representative said 'it was unbelievable that the EU would serve as a broker for Spanish pirates who have already decimated other flatfish stocks around the world'.[20] The EU then set its own quota for turbot in the high seas off Canada's EEZ which was much higher than NAFO's. Tobin said that Bonino would face arrest 'if she's on the back of one of those trawlers'.[21]

Canada passed a law allowing Canadian officials to arrest vessels which ignored NAFO's conservation measures on the high seas. Such a law had no international standing. Anyone could exercise the right to fish the high seas. The only qualification came if the flag state of the fishing vessel had signed up to regional agreements. While the EU accepted NAFO they regarded its quotas as unfair and so would not act to curtail the Spanish or Portuguese fleets.

Canada then passed further legislation allowing them to board and seize vessels in the high seas which they suspected of illegal fishing. This too had no international standing. Spain threatened to send a naval vessel to protect their fleet. Negotiations failed and the Canadian Prime Minister Jean Chrétien gave Tobin permission to arrange the seizure of Spanish ships that were violating conservation practices. Thus it came about that in 1995 a Canadian patrol vessel fired shots across the bow of the Spanish trawler *Estai* in the high seas off Canada's EEZ. The Canadians threatened to fire on the ship if Captain Gonzales on the *Estai* did not stop the ship. He did and the *Estai* was arrested. Captain Gonzales was charged with unlawful fishing and throwing fishing equipment overboard. When this was recovered it was found that the mesh in the Spanish net was below the minimum size set by NAFO, allowing the fishers to haul in illegally small fish. Tobin called the net an 'ecological monstrosity' and a 'weapon of destruction'.[22] Most of the *Estai*'s catch was undersized and its records were falsified. Tobin claimed that the Spanish fishermen were 'rogue pirates'.[23] Nevertheless the UN condemned Canada's move and sent a patrol vessel. A

Spanish official called the arrest of the *Estai* 'an act of war against a sovereign country',[24] and the *Estai* fishers pelted the Canadian Embassy in Madrid with fish. Bonino said that Tobin was the real pirate. Canada's action was very popular at home and Tobin received a standing ovation in parliament. According to Raymond Blake, in 'Water Buoys the Nation':

> Brian Tobin did for fish what no other person has done since John Cabot: he brought it to the attention of the world, and for Canadians he made them realize that the fish off their shore was theirs ... [He] made them realize that the fish that swim in the waters on the continental shelf off the east coast are Canadian and Canada needs to protect them.[25]

Tobin defended the arrest at a UN forum arguing that coastal states needed to take responsibility for conservation beyond the 200 mile limit when fish crossed in and out from this border, and then added on a more passionate note: 'we're down to the last, lonely, unloved, unattractive little turbot, clinging by its fingernails to the Grand Banks of Newfoundland, saying "someone reach out and save me, this eleventh hour as I'm about to go down to extinction"'.[26]

The owners of the *Estai* posted a $500,000 bond and the ship was released. After an agreement was reached between Canada and Spain covering conservation and enforcement the bond was returned.[27] Finally in 1995 the UN brought in the Straddling Stocks Agreement.[28] When ratified this Agreement allowed coastal states to seize ships fishing illegally on the high seas (in the sense of going against regional agreements) when the flag state does not take enforcement action against its vessels. The Agreement was ratified in 1999 and by 2009, 75 countries had signed up (as compared to 158 signatories to the Law of the Sea). The number of signatories is disappointingly low. However Spain and Portugal ratified the Agreement in 2003. Non-compliance by EU vessels off Canada's EEZ continues to be a problem. Some members of the EU are reluctant to prosecute clear violations of NAFO rules but the current situation is an improvement on 1995. The Straddling Stocks Agreement is the first major change in the legal regime over the high seas since Grotius.

Both Iceland and Canada tried to get assistance from the international community when they saw their fish populations were in danger. When these efforts failed they acted on their own to expand control over the oceans, Iceland by extending its exclusive fishing zone to 200 nautical miles and Canada by acting as an enforcer of conservation measures even beyond that zone. Both countries argued that the international instruments in the form of the Law of the Sea, the International Court of Justice or NAFO were acting too slowly. These bodies have now swung behind the Icelandic and Canadian decisions.

The Icelandic and Canadian fishing histories mentioned here seem to point to the conclusion that having sovereign rights over 200 nautical miles is highly desirable both for the coastal state and the fish populations. This is clearly true for Iceland. However the Canadian case shows us that even when the coastal state has sovereign rights and the power to enforce fishing regulations fish populations can plummet. When foreign fishers were evicted from the 200 mile zone fish numbers continued to fall as local fishers took up the hunt more aggressively. Many other states that have adequate policing powers have failed to protect their fisheries. I visited the Lofoten Islands in Norway in 2006. These islands had been thriving fishing areas. In 2006 I saw abandoned villages, and one quite large village called Å where the total area has become a museum with the castor oil factory, dried fish plant and boat building workshops, as well as shops and houses, preserved for visitors to look at, a poignant reminder of a way of life now gone. In the Lofotens there is currently a very small fleet, fishing for a very limited season on very low quotas. Fishermen told me that they had to go out under-crewed to make the venture worth their while.

When coastal states have inadequate policing the 200 mile zone cannot be protected. As noted above, overfishing in Somalia by foreign fishers was the main reason for the rise in piracy. West Africa and Southeast Asia have also had problems in enforcing sovereign rights over the EEZs. The Canadian example revealed the problem of protecting fish that swim in and out of the EEZ from the high seas. The Straddling Stocks Agreement is helpful for the Canadians to give them a right to pursue illegal fishers beyond their EEZ. Many other states, however, would not have the ability to take on such a role. From these considerations alone it seems desirable to place the governance of the fishing industry beyond territorial seas, in the hands of an international body that has access to the best science for management and the power to enforce regulations that flow out of that science. Crucially too, a co-operative relationship with fishers has the best chance of good outcomes. The property rights model adopted by Iceland could have general application and help ensure such co-operation. This change would mean that coastal states relinquish their rights to the EEZ beyond the territorial sea. Another line of argument that leads into the same conclusion comes from looking at what is happening on the high seas. One illustration is fish 'piracy'.

FISH PIRACY

As commercial fishing operations in the Northern Hemisphere are wound down because of too few fish, many operators have moved into the Southern Ocean. This is the domain of a fish so commercially valuable it is called 'white gold': the Patagonian toothfish (PTF) (*Dissostichus eleginoides*), also known as mero, Chilean sea bass, loup du chili, Bacalao de profundidad, Butterfish and even 'the sweetheart fish of the 90s' because of its desirability as a food source.[29] These fish are chased by legal and illegal fishers in the high seas and in the EEZs of remote islands belonging to Australia (Heard and Macdonald), France (Kerguelen and others) and South Africa. Large-scale fishing of the PTF began only in the early 1990s, yet with the efficiency of contemporary fishing vessels the population is now under threat[30] and the conservation organisations Greenpeace and Ocean Defence are calling for a moratorium.

The PTF are migratory and may swim within the 200 mile zone or the high seas. The nations that were primarily interested in taking PTF formed an alliance – the Commission for the Conservation of Antarctic Marine Living Resources (CCAMLR) – and drew up an agreement to manage the fishery, monitor the ecosystem and reduce seabird bycatch. The management is based on catch limits and vessel limits.[31] In 2008 there were 24 member states in the alliance.

The 'piracy' label is applied to fishers who are violating the conservation agreement. They are 'fishery pirates' or in Spanish, *pescadores furtivos*. The 'pirates' are usually not from states that are members of CCAMLR though member violations have occurred. Both Spain and Russia are members yet they have been implicated in illegal fishing. An example follows below. The 'pirates' often use flags of convenience. They buy the right to register a fishing vessel to a particular nation such as Belize, Panama, Liberia, Cyprus or Singapore and to fly the flag of that country.[32] These states are not Parties to CCAMLR. Some estimate that the number of PTF caught by the 'pirates' could be 50 per cent of the total catch.[33]

In a recent attempt to combat the fishery piracy problem another organisation has been formed. This is the Coalition of Legal

Toothfish Operators (COLTO).[34] They look out for and report illegal fishers. These operators are from France, the Falkland Islands, South Africa, Japan, Spain, New Zealand, Australia, Argentina and Chile. Pescanova, a large Spanish fishing company, is in COLTO, and yet it is well known that Pescanova has been involved in illegal fishing in the CCAMLR area – for example, in 1997 *Magallanes I* was impounded for illegal fishing off Kerguelen Island. Pescanova was the owner of *Magallanes I*.[35] This leads to a suspicion that at least some members of COLTO are there to protect business interests more than a legal fishery.

If the nations who register the fishing vessels of the 'pirates' are not party to CCAMLR or other high seas fishing agreements then the fishers are not strictly doing anything illegal (hence the quotation marks around the word 'pirates') if these 'pirates' keep to the high seas. The Straddling Stocks Agreement doesn't apply in the Southern Ocean for reasons given below. If 'pirate' ships are caught fishing in an EEZ of a foreign country then legal penalties can apply, but catching such a ship can be treacherous in this ocean with big waves, strong winds and icy conditions, and top legal teams may be brought in to defend against the charges as large amounts of money are at stake.

In 2003 an Australian patrol boat, the *Southern Supporter*, chased a suspicious trawler, the *Viarsa 1*, for 2,200 nautical miles over a three-week period. With the help of South African and British ships, the *Viarsa 1* was arrested and escorted back to Australia. In the ensuing court case, the crew were acquitted in a jury trial even though the Australian Government seemed to be in little doubt that they had engaged in illegal fishing. The international nature of these operations is illustrated here. The *Viarsa 1* was registered in Uruguay (although Uruguay deregistered it after the capture), the skipper was Uruguayan, and the crew were Spanish and Chilean. The Spanish operations manager chartered the vessel from a Panamanian company. Viarsa Fishing Company is Spanish and the owner of the *Viarsa 1* is Spanish.[36]

Spanish, Ukrainian and Russian involvement in illegal fishing has been well documented.[37] In a particularly dramatic case involving a Spanish captain and Russian ships in 2001, two ships were brought to the Australian mainland and the crews put on trial. The *Southern Supporter* noticed the Russian fishing boat, *Lena*, close to Heard Island. When asked to proceed to Fremantle on the Australian mainland, Jose Sanchez, the Spanish captain of the *Lena*, said that she was merely in transit and cheekily took on fuel from another 'pirate' vessel, *Florens I*. The *Southern Supporter* was then lured away by a fake distress call and the *Lena* escaped. She was repainted to change her appearance.[38]

Australia then sent out two naval vessels, the HMAS *Canberra* and the *Westralia*. After catching up with the *Lena*, four armed men were lowered onto her deck by a helicopter from the *Canberra*. An illegal toothfish catch was found on board and the *Lena* was taken back to Australia. A month later *Canberra*'s helicopter spotted the *Volga*, another Russian fishing vessel near Heard Island. She was also boarded by the Australian navy. She was carrying 127 tons of toothfish worth about $1,500,000. The *Volga* was taken to Fremantle and the crew charged with illegal fishing. Some of the crew on both boats received heavy fines. The *Lena* was scuttled, and a $4,177,500 bond was placed on *Volga* for its release. The Russian owners appealed to the International Tribunal for the Law of the Sea against the severity of this decision. They lost.[39]

Some of the ramifications of the impacts of the 'piracy' of the PTF are clearly evident or easy to predict. The PTF grows slowly to more than two metres long. It can live for 50 years and does not breed until it is at least ten years old. Taking out the younger fish will mean that the breeding cycle stalls. There are effects on other species too. The PTF is eaten by sperm whale and 'scientists estimate that for elephant seals of the sub-Antarctic Heard Island, toothfish comprises a major part of their fish diet'.[40]

Albatross and other seabirds are caught in the common long-line fishing method. The standard fishing techniques for the PTF involve the use of long lines with baited hooks, strung out for kilometres across the ocean. A single boat can set long lines with as many as 50,000 hooks every day.[41] Albatross and petrels take the bait and get caught in the hooks that are subsequently sunk. When this happens, the birds drown. Long-line operations have been identified as the single most important factor in the current global decline of albatross populations, especially in the Southern Ocean,[42] and the albatross are currently the most threatened and vulnerable group of all seabirds.[43]

Sir David Attenborough comments on the effects of long-line fishing on albatross:

When mankind first entered their domain, as little as 500 years ago, albatrosses had already been masters of the oceans for 50 million years. But, our fleeting contact with these gentle giants has sent 19 of the 21 albatross species soaring towards extinction.

For the want of a few inexpensive changes to fishing methods, an albatross drowns on the end of a fishing hook every five minutes.[44]

There are varying estimates of the status of the PTF populations but a recent report indicates that current fishing practices could devastate the population for decades.[45] In such a situation it would be prudent to take a precautionary approach. CCAMLR's *aim* has been to adopt a precautionary approach, namely to collect data, acknowledge the uncertainties therein, but to make management decisions bearing in mind and in a manner that strives to minimise the risk of long-term adverse effects on the environment, rather than delaying decisions until all necessary data are available. CCAMLR has been a world leader in adopting this approach.

CCAMLR came into force in 1982 but the crisis surrounding the PTF is of fairly recent origin. Although expressing the precautionary principle, CCAMLR has been accused of acting too slowly and is now itself in crisis.[46] Bederman summarises the problems as follows:

Decision-making on matters of substance is by consensus, which means that a single State can block a conservation initiative. More seriously, the Commission's decisions are treated as recommendations and do not have binding effect on member States in the face of an explicit objection. ... The Commission lacks the ability to ... acquire and develop scientific data that is essential to the sound management of Southern Ocean fisheries. Finally, CCAMLR utterly lacks any means to enforce conservation measures as against Parties or non-Parties.[47]

So, in effect, a marine organisation that has tried more than most 'to deploy great caution' only stumbles. One could plead inadequate institutional arrangements or lack of funding but the core problem may lie elsewhere. The stated aim of the Convention is 'to conserve marine life of the Southern Ocean. However this does not exclude harvesting carried out in a rational manner.'[48] It seems that the conservation measures are increasingly geared primarily towards economic outcomes that are the dominant interests of the participants.[49] So if Parties or non-Parties do not see an economic advantage in limiting catches they may not be inclined to do so. 'Pirate' ships flying flags of convenience are only subject to regulation by their flag state, which may be uninterested in enforcing fisheries regulations at the risk of jeopardising licence revenue. It is likely too that this perception of the Convention as a tool for economic gain has led some non-Parties to ignore its rules, a point which will be taken up again below.

When CCAMLR was set up it attempted to adopt an ecosystem approach, but rather than conduct the essential baseline ecosystem studies it focused on the plight of individual species such as krill.[50] Since 1986 there has been a more rigorous attempt to introduce an ecosystem management approach. CCAMLR has been described as the first major international fishing regime to focus on ecosystems rather than species.[51] Bederman, writing in 2000, praises CCAMLR for 'featuring the first and strongest ecosystem management approach in a regional fisheries treaty',[52] but the institutional inadequacies in CCAMLR summarised above have hampered its effectiveness.

One major problem is the inability of CCAMLR to control 'piracy', the unsustainable practices of the 'pirates' and the heavy reliance on long-line fishing. Some CCAMLR members have changed their fishing practices out of concern for the ecosystem impact, but the slaughter continues in most areas of the Southern Ocean.[53] The second reason why CCAMLR has been unsuccessful in protecting ecosystems may have to do with the orientation of conservation measures in the Law of the Sea which are directed very much towards species rather than ecosystems. CCAMLR is a separate treaty from the Law of the Sea but the latter still operates over the high seas in the Southern Ocean.

Roberts argues that the way to protect ecosystems is to set aside special conservation zones or reserves.[54] A system of reserves would greatly benefit non-migratory fish. However, it is inadequate for fish that traverse vast distances in the ocean, such as the PTF. If the fish migrate they may be targeted out of conservation zones with degrading effects on the broad ecosystem. The albatross also roam over the high seas. Isolating a 'conservation zone' may not assist their plight. Conservation zones take on particular relevance when habitat preservation is a high priority. In the relatively pristine waters of the Southern Ocean this is not yet a primary concern. The key issues are the ramifications of 'pirate' fishing practices for birds, seals and whales, as well as their impact on the PTF. In addition there could also be unpredictable ecosystem impacts resulting from the decimation of the PTF stocks. As Heywood states: 'any use of biological resources results, to some degree, in alteration of ecosystems, and often in their simplification, an effect which may result in ecosystem instability'.[55]

There have been more and more serious attempts by CCAMLR members individually or in co-operation to enforce conservation measures within their EEZs in the Southern Ocean. Australian laws have been strengthened to allow greater surveillance and to give fisheries, environment and law enforcement agencies added powers to share information. These changes to the Fisheries Management Act introduced in 2007 also make it easier to impose harsher penalties, including the confiscation of foreign boats, fishing gear and any fish catch if the crew are convicted of illegal fishing. Greater use of arms by the navy is now allowed in the apprehension of suspected fishery 'pirates'. Patrol vessels are able to fire directly to disable a vessel which is trying to escape.[56]

Fines have been steadily increasing through the French courts for illegal fishing off Kerguelen, from 15,000 francs in 1996 to 400,000 and then 3,000,000 in 1997 – for different amounts of fish admittedly, but the trend is still upwards. The bond for the return of the *Volga* mentioned above was $4,177,500, set by an Australian court. A Chilean skipper was fined 2 million pounds sterling in a Falkland court for illegal fishing and he lost the catch valued at 7 million pounds as a result of the case. The owners were fined 60 million pounds and refused to pay, so the ship was scuttled at Shag Rocks near Lively Island in the Falklands.[57]

For some years CCAMLR states have offered assistance to each other in apprehending fishery pirates. Close ties are developing between the French and Australian navies, who began doing joint training together in 2006, focusing especially on the tricky manoeuvre of boarding a hostile ship. France and Australia have also decided to pool resources in research and surveillance activity in their respective EEZs in the Southern Ocean. Such co-operation has followed on from greater links between the fishery pirates who communicate with each other using satellite phones to inform the fleets about the location of patrol vessels. They offer assistance to each other with re-fuelling on the high seas and collude on diversionary tactics to mislead patrols.

One deterrent for the fishery pirates is to make it difficult for them to land illegally caught fish. Given that the PTF is so popular, finding ports that will accept the catches has been easy. However, Australia's new fisheries legislation introduces a licence for fish buyers who will be required to report on the fish delivered to them. In the first case to be brought under similar legislation in the US in 2006, a Spanish businessman, Antonio Vidal, was fined $400,000 for illegally importing PTF and a Ukrainian company also involved in this import was fined $100,000.[58]

One of the major factors behind over-harvesting is the unwillingness or inability of nations either individually or collectively to take responsibility for fishing *on the high seas*. As a result, commercial interests dominate and there seems to be little concern for taking only a small portion or for sustainable practices even to preserve the species never mind the ecosystem. The 'pirates' are after quick profits and will move to new fishing grounds when old ones become barren. This situation would be for Hardin a predictable consequence of treating the sea as commons. Arguing about the commons, land freely open to any user, Hardin said that if the users were selfish and the exploitation of the resource exceeded the natural rate of replenishment of the resource then the tragedy of the commons would result.[59]

The Southern Ocean forms a vast marine commons. Although larger and larger tracts of sea are coming under the control of coastal states through the EEZ in the Law of the Sea, the high seas are still unowned territory and the right to fish the high seas by anyone is enshrined in the Law of the Sea.[60] There are examples, such as with the Icelandic cod fishery, where over-harvesting has been turned around when communities get together and agree to limit their take.

The move to extend the jurisdiction of coastal states to 200 nautical miles allows the possibility that such states will take responsibility for over-harvesting in those waters. Indeed this is what Australia is attempting to do in its sub-Antarctic zones around Heard and McDonald Islands.[61] Some coastal states have usurped the right to manage fisheries beyond 200 miles in response to the damage caused to them by fishing practices permitted by the freedom of the seas regime. The Canadian Government justified this initiative by arguing that 'due to the biological unity of straddling stocks, overfishing them beyond the 200 mile limit will also deplete them within the zone under national jurisdiction'.[62]

However even the extension of coastal jurisdiction to 200 nautical miles enshrined in international law may be seen as unjust. It is then open to the coastal state to take exclusive economic advantage of this zone. Australia's maritime zone now extends to 8 million square kilometres.[63] Why should coastal states be advantaged in this way given that fishing in distant waters does not require a close home port or any port within the state? There could be here another incentive for 'piracy' to continue when the legal carving up of the oceans is seen as advantageous to some over others.

The PTF move into the EEZs at times, but spend a great deal of their life in the high seas. A zoning system for the oceans seems to be an unimaginative extension of land-based systems. The breakthrough in international jurisprudence that came with the Straddling Stocks Agreement has helped the Canadian turbot. It offers some protection for fish that move in and out of the high seas. It promotes the interests of the coastal states in relation to the management of high-seas fisheries adjacent to their economic zones in that the coastal states can enforce fishing regulations when the flag states don't act. However the Southern Ocean lacks coastal states and hence the UN Agreement does not apply. (Australia and France, for instance, have island territories in the Southern Oceans with EEZs around them but they don't count as coastal states.) The Southern Ocean remains a unique form of global commons and may well be the last such area on earth.[64]

Pureza makes the general claim that 'only economic, technological and militarily powerful countries have effectively benefited from the open access opportunities, since the regulatory minimalism of the freedom of the seas doctrine has led to 'the cynicism of the "first come, first served" rule'.[65] Perhaps it is time to re-visit the doctrine of the freedom of the high seas but in a way that will ensure equity.

The Law of the Sea may be another part of the problem, unwittingly promoting exploitation rather than putting a curb on it. This follows for several reasons:

1. Although the Law of the Sea addresses issues of conservation it is not binding on states that don't ratify the Law.
2. Conservation is discussed as conservation of a resource rather than an ecosystem, as mentioned above.[66]
3. Conservation is tied up with maximum sustainable yield as laid down in Article 119: 'States shall ... take measures which are designed, on the best scientific evidence available to the States concerned, to maintain or restore populations of harvested species at levels which can produce the maximum sustainable yield, as qualified by relevant environmental and economic factors.' The aim is to secure the maximum supply of food and other marine products.
4. The Law of the Sea Article 116 enshrines the concept of the freedom to fish the high seas.
5. In the way zoning systems have been worked out an inequitable arrangement has been placed into law. The EEZs and their extensions deliver unfair advantages to coastal states or territories. This could lead to resentment and might lie behind some of the 'pirate' operations.

Nothing short of a radical change in ocean governance, perhaps along the lines developed in Chapter 9 below, is necessary to tackle problems such as fish piracy.

THREATS TO FISH POPULATIONS FROM CLIMATE CHANGE AND OCEAN ACIDIFICATION

Fish in the polar regions will experience a loss of food sources from warming seas and ocean acidification, as mentioned in Chapter 2, where I also outlined the long-term consequences of increased CO_2 up-take in the oceans: mass extinctions of fish. Ocean acidification can also have a devastating effect on cold water and warm water corals destroying habitats and food sources for many fish populations. Corals die-off from bleaching will compound this problem. A report put out by the UN Environmental Program in 2008, focusing on 'the economically important 10–15 per cent of the oceans and seas where fish stocks have been and remain concentrated', states that more than 80 per cent of the coral reefs in these areas may be lost due to rising sea temperatures and acidification.[67] Optimistic projections date this likely impact at 2080.[68]

Coral bleaching is already evident in most tropical coral reefs with a large proportion dying. A rise of a mere 1° C above the usual monthly average in summer is enough to produce this effect. As noted in Chapter 2, early signs of damage caused by acidification have been detected recently, informing the view that severe impacts may be only decades away rather than centuries, as previously believed.[69]

There are also growing fears about changed ocean circulation patterns in a warmer sea. The 'turning over' of water in most of the world's fishing grounds plays a crucial role in flushing away pollutants and in providing fish with food from the ocean depths. This mechanism is temperature related. The 'cooler and heavier seawater sinks into the deep sea' but if the water doesn't cool then there will be a reduction in 'the intensity and frequency of coastal flushing mechanisms, particularly at lower to medium latitudes over the next 100 years which in turn will impact on both nutrient and larval transport and increase the risk of pollution and dead zones'.[70]

THE WAR ON FISH

So far I have focused on the idea of fish as a resource. This is the way we are encouraged to think about fish when reading the Law of the Sea and the various agreements mentioned above. Yet what if fish can feel pain and suffer in degraded environments? Should these factors be taken into account? Do humans have a moral duty to consider the death and suffering of fish? Is there a moral issue in the extinction of a marine species?

Human's moral relations with whales and dolphins are fairly commonly accepted. The moratorium on whaling now in place is motivated at least in part by the thought that humans have no moral right to kill whales. A moral duty not to harm fish was until recently dismissed as it was thought that fish do not feel pain, so they couldn't be harmed by what we do to them. A carefully conducted study at the University of Edinburgh in 2003 pointed to the conclusion that fish do indeed feel pain. Sneddon and colleagues distinguished between the simple detection of and reflex response to a noxious stimuli, and pain perception:

> To demonstrate that an animal is capable of pain perception, it must be shown that, first, the animal can perceive the adverse sensory stimulus and second, that it reacts both physiologically (e.g. inflammation, cardio-vascular changes) and behaviourally (moves away from the stimulus) ... it is necessary to show that the animal learns that the stimulus is associated with unpleasant experience and avoids it.[71]

The researchers cite studies showing that fish learn to avoid electric shocks and hooking during angling. This study used trout and the injection of noxious substances into their lips. Receptors for noxious substances were found. Many were stimulated at a lower level than human skin indicating a more sensitive response closer to the response of a mammalian eye.[72] The ventilation rate rose to a rate similar to swimming at their maximum speed. Ventilation rate also increases in humans enduring noxious stimulation. The injected trout 'performed anomalous behaviours' such as rocking. 'This is reminiscent of the stereotypical rocking behaviour of primates that is believed to be an indicator of poor welfare.'[73] The trout rubbed their lips against hard surfaces in the tanks as humans might rub an injured area to ameliorate the intensity of pain. Reflexes cannot explain these results. Rather, the authors conclude, the fish feel pain.[74] If they feel pain in these circumstances it is reasonable to conjecture that they have the capacity to feel pain in many other noxious situations, for example when suffocating after capture.

In 1789 the English philosopher Jeremy Bentham argued that the important question for morality is whether the individual (human or otherwise) can suffer. If a creature can suffer then humans have a moral duty not to bring that about.[75] What happens if we draw out the consequences of taking Bentham's view seriously along with the idea that fish can indeed suffer? If fish do feel pain then, following Bentham, their interests should be taken into account. Humans have conflicting interests however, since eating fish satisfies an important part of our diet, the protein intake. This could be met in other ways but they are not always available and also have moral implications, for example, eating meat goes against the interests of animals. Vegetarian options, though less compromising, require land clearing and displacement of animals. A morally defensible position might allow the capture of fish for food so long as their suffering is minimised by a quick death. It might also require of humans that we refrain from acting in a way that pollutes the marine environment, thus giving rise to suffering in fish. Species survival is a moral issue if the world becomes impoverished for future generations of humans who might want to eat fish, or enjoy the fact that wild creatures still exist.

This line of reasoning is quite alien to contemporary practice. It is as if humans are waging a war on fish. More efficient boats and equipment mean that it is now quite easy to work out where fish congregate and to go after them. No thought is given to the suffering of fish once captured. If fish suffer from pollutants this is also rarely seen as a moral problem; the only question asked is: Will we still be able to eat them? Species survival is usually only considered, if at all, as an issue from the perspective of fish as a resource to humans, not in terms of species extinction as a moral issue.

Around Southeast Asia, Oceania, East Africa and Crete, fish have been dynamited out of the water or poisoned with cyanide to make capture easy. Since 9/11 dynamite has been in short supply. However cyanide is still in use along with other poisons. Needless, to say these areas are experiencing drastic declines in fish populations. Some scientists predict the total destruction of all coral reefs in the Southeast Asian region by 2020 if these practices continue.[76] All governments in these regions have laws that declare fishing with both cyanide and dynamite illegal, but they are usually not enforced because of lack of resources. The depletion of fish in one place leads to poaching of fish from other countries and consequent political tension. It is sometimes foreign fishers who conduct dynamite fishing, for example, Vietnamese and Chinese in the Philippines.[77]

The ruthless plunder of fish and the use of explosives and poison against them supports the idea that a war is being waged against fish, a war which humans are likely to win.[78]

Yet we will suffer from that victory. The UN Food and Agriculture Organisation, Fisheries and Aquaculture Department (FAO) put out a report every two years on the State of the World Fisheries. In the 2006 report, 75 per cent of the world commercial fish populations are documented as depleted, recovering from depletion, fully exploited or over exploited.[79] In 2007 a WorldWatch Institute study found that three quarters of the world's fish stock had been over exploited.[80] According to the research conducted by Meryl Williams, 'Nearly 100 million people are directly dependent on the fishing industries and their related service sectors in Southeast Asia, and nearly all Southeast Asians are fish consumers.'[81] The consequences of fish population collapses are unimaginable.

How would a moral position in relation to fish work in practice? It is impossible to employ surveillance measures on all fishing vessels and a change in fishing culture will probably only come from valuing fish as living sentient beings rather than simply a food source. It is only possible to get to that point with changes of thinking in cultures more generally. This is happening slowly in relation to animals. It should be easier to bring in pollution controls as these can be government sponsored and partially land-based. There may be other advantages apart from helping the fish for example, opening up areas to marine tourism. The argument for species protection based on intergenerational ethics hinges on present-day humans seeing the justice of this appeal. It is sad that it has taken the reality of climate change to get people thinking that future generations matter and that the type of world they will inherit matters.

To return to the fish, if they are creatures who can suffer and thus have an interest in living in an environment which minimises their suffering then it is not a big leap to say they should be accorded some rights to their habitat, the ocean. Whoever does own the oceans should take account of the moral claims that fish have on us. Who will protect the fish if no one owns the ocean? Responsible measures haven't usually been employed in the EEZs, fish suffer, and species die. How much worse off are the fish in the high seas? The FAO Report mentioned above states that the fish populations in the high seas are the most over exploited. This is a particular problem with highly migratory oceanic sharks and fish that swim in and out of the high seas, the 'straddling stocks'.[82] Some very damaging and cruel fishing practices do occur close to shore, but it is probably not realistic to think that states will ever give up ownership of sea territory out to 12 nautical miles. However the 200 nautical mile jurisdiction that states 'enjoy' must also partly be perceived as a burden. Poorer countries in particular have little hope of promoting sustainable fishing practices or conservation measures in this zone. Without change, these parts of the sea may be biologically dead in the near future. Thus the handing over of ownership to an international body beyond the 12 nautical miles zone may not be seen as such a big step. The coastal state would be giving up exclusive economic exploitation of the EEZ but could still be involved in that exploitation in a sustainable and equitable way. If the international body had enough resources, and I will suggest below where they might come from, then it could act to prevent the demise of fish populations and implement other measures to relieve the suffering of fish. Roberts argues passionately for the setting up of oceanic reserves to allow recovery of fish populations. In 2007 he noted that 'Twelve per cent of the world's land is now contained in protected areas, whereas the corresponding figure for the sea is but three-fifths of one per cent. Worse still, most marine protected areas allow fishing to continue.'[83] Roberts claims that in order to have a significant effect, reserves would need to cover 30 per cent of the sea with many located in the high seas.[84] Yet he fails to suggest any mechanisms for setting these up. If, as Roberts asserts, marine scientists are swinging behind the idea of an extensive system of oceanic reserves, then an international body will need to be involved. This is another reason why ownership of the seas beyond 12 nautical miles should be taken up by some such body. Whereof more anon.

6
Cetaceans and the Sea

Though the search for extraterrestrial intelligence may take a very long time, we could not do better than to start with a program of rehumanisation by making friends with the whales and the dolphins. ... They have behaved benignly and in many cases affectionately towards us. We have systematically slaughtered them.

Carl Sagan, *The Cosmic Connection*[1]

WHALES AND DOLPHINS

In 1977 two captive female bottlenose dolphins held in a university-supported marine laboratory in Hawaii were lifted from their tanks and turned loose into the nearby Pacific Ocean. The two men who released the dolphins lived and worked in the marine laboratory with the the director Dr Louis Herman assisting in the behavioural experiments with the dolphins. These researchers became disenchanted with the idea of using captive dolphins for scientific experiments. In a press conference after the release they argued that what they did was not a crime. Rather it was a crime to keep 'dolphins – intelligent, highly aware creatures, with no criminal record of their own – in solitary confinement, in small concrete tanks, made to do repetitious experiments for life'.[2]

Following on from the ethical question I raised about fish, we can ask, do dolphins have a moral claim on us such that there are some things which we ought not to do to them and some things which we ought to do to enhance their flourishing? Given that dolphins normally live in the ocean would an ethical stance in relation to them bear on ocean policy? Is our duty to dolphins simply a matter of leaving them alone or are there threats to their well-being that call for human intervention?

Dolphins fall into the order of Cetacea which also includes whales and porpoises. Living cetaceans are divided into two major groups: Mysticeti or baleen whales and Odontoceti or toothed whales. The former use baleen, sometimes called whalebone, to filter out small planktonic organisms for their food intake. Baleen whales are known as the 'great whales' and the largest, the blue whale, is the largest of all living animals and also the largest animal ever known to have existed, greatly exceeding the size of the largest dinosaurs. This is curious given that their diet consists of very small organisms.[3] Baleen whales consist of the right whales (which include bowheads), the pigmy right whale, the narwhal, the beluga, the grey whale and rorquals. Rorquals include the fin whale, Bryde's whale, sei whale, blue whale, minke whale, and humpback whales. Toothed whales select and capture individual prey such as fish or squid, rather than filter out small organisms. Under the toothed whales we find the following families: sperm whales, white whales, beaked whales, dolphins and small-toothed whales (including the killer whale or orca), porpoises including the vaquita, river dolphins, the Amazon River dolphin, Chinese river dolphin and the franciscana.[4]

One feature of some of the whales is of particular importance when thinking about risks to their well-being, and that is their migration patterns. Figure 6.1 shows the migration routes of the humpbacks.

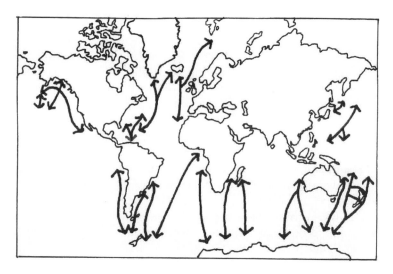

Figure 6.1   Migration of humpback whales. Modified from R. Harrison and M. M. Bryden (eds), *Whales, Dolphins and Porpoises*, Sydney: Golden Press Pty Ltd, 1989, p. 98.

Baleen whales migrate over huge distances from food sites to reproduction sites. Humpback and grey whales regularly migrate close to the coastlines of continents. Some whales do not have fixed migration patterns but travel into different areas according to the environmental conditions. This is particularly so for the Bryde's whale, sei whale and bowhead whale. Sperm whales also migrate over large distances in complex arrangements related to age and sex groupings.[5]

CETACEANS AND MORALITY

Given that cetaceans are more complex than fish it is possible to include them within various moral positions, for instance the animal rights perspectives developed by Tom Regan and Ted Benton or the utilitarian approach of Peter Singer. Singer expands on Bentham's view, and I have followed through the implications for cetaceans in another place.[6] Singer asserts that humans and non-human animals are equal. What does that mean? He rejects the idea that we must treat every animal in the same way but he supports the equal consideration of animals. Equal consideration for different beings may lead to different treatment. It will depend on the nature of the animals being considered. 'Concern for the well being of a child growing up in America would require that we teach him to read; a concern for the well-being of a pig may require no more than that we leave him alone with other pigs in a place where there is adequate food and room to run freely.'[7]

Singer believes the extension is justified because non-human animals have needs and interests just like humans. Behind the having of needs and interests is the capacity to suffer, or to experience pleasure. Singer believes that if a being has no capacity to suffer or to enjoy then it has no interests. If a being suffers there can be no moral justification for refusing to take that suffering into consideration and it should be considered equally in the sense that the suffering of one being should be counted equally with the suffering of another being, although he acknowledges that the comparisons might sometimes be quite rough. If a being isn't capable of suffering or of experiencing enjoyment there is nothing to be taken into account. It is 30 years since Singer presented this view for the first time. However his position is still influential in moral debates.[8]

The capacity of whales to suffer cannot be seriously doubted today. D'Amato and Chopra write:

> When whales are harpooned and dying, their characteristic whistles change dramatically to a low monotone. In contrast, in the normal healthy state, their whistles are beautiful birdlike sounds with trills and arpeggios, glissandos and sitar-like bends in the notes. This change is clearly analogous to the transformation in human expression from talking (or singing) in the

normal state to crying when in pain. Additionally, there can be little physiological doubt that whales feel pain; indeed, the real question is whether they perceive acute pain to an even greater degree than humans. This latter possibility is evidenced by the far wider range of skin sensations apparently registered by the complex cerebral cortex of the whale.[9]

Extending equal consideration to cetaceans would entail the cessation of whaling and the removal as far as possible of other threats to their life or well-being.

There is another moral position which appeals to intergenerational ethics. It contains the idea that as a matter of justice we should not leave a degraded environment for future generations to redress. Given that cetaceans add value to the world because of their intrinsic worth, because of their unique role in the ecosystem and because of the joy and inspiration they give to humans, then humans have a moral duty to preserve whales for future generations.

THREATS FACING CETACEANS

1. Slaughter

Whaling kills whales slowly, causing great suffering. In 1986 a moratorium on commercial whaling was passed through the International Whaling Commission (IWC). However, countries can still apply for a special permit to conduct scientific whaling. Since the moratorium on commercial whaling, Japan, Norway and Iceland have been granted these permits. Norway stopped its scientific whaling programme in 1995 and resumed commercial whaling, taking hundreds of minke whales a year increasing to a quota of 1052 in 2008. The quota for 2009 dropped to 885 minke whales. This was attributed to consumers' growing disinterest in whale meat.[10] Support for whaling amongst younger Norwegians is also in decline.[11] Iceland killed about 100 fin whales a year from 1986 to 1989 under scientific permits, and has had permission to kill about 40 minke whales a year since 2003 under the scientific permit. Iceland resumed commercial whaling in 2006, stopped in 2007, and issued quotas in 2009 for 100 minke whales and 150 fin whales. The IWC rules allow signatories to put in an objection to a policy such as the moratorium on commercial whaling and then proceed to act against that policy.

Japan expanded its government-subsidised scientific whaling program from 273 minke whales in 1986–7 to 935 in 2007–8. The 2007–8 permit included 50 humpback and 50 fin whales. However, during the 2007–8 season anti-whaling activists on the *Sea Shephard* made contact with the whaling fleet, disrupted whaling operations and created bad publicity for the whalers. As a result no fin or humpback whales were taken and only half the minke quota. In 2008–9, the quota for minke whales dropped to 750, with 50 fin whales but no humpbacks. This may be attributable to a decline in demand. *Yushin*, a major whale meat shop and restaurant in Tokyo, announced that it will close in 2010. Apparently there have also been difficulties in finding Japanese crew.[12] Senior officials in Japan are openingly questioning the continuation of whaling, pointing out the damage to Japan's image in the English-speaking world and the fact that it is such a small part of the economy, 'less than one-tenth the value of the country's annual market for toothbrushes'.[13] Japanese groups opposed to whaling are also on the rise.[14]

The scientific permits are granted by the IWC but it is the member nation conducting the research that has the ultimate responsibility for issuing the permit. For years the IWC passed a number of resolutions asking governments to refrain from issuing these permits. There are two main objections from the Scientific Committee of the IWC. Firstly, most of Japan's slaughter of whales is in the Southern Ocean Sanctuary, the marine area surrounding the Antarctic continent where commercial whaling is banned by the IWC. One of the stated aims of the Japanese research is to 'improve management', namely, to work out how many whales can be killed without threatening the species. However if the whales are in the Sanctuary this information is unnecessary as the killing of these whales is prohibited (except for the scientific loophole). Secondly the Scientific Committee is not persuaded by the merits of Japan's research. The IWC passed a resolution strongly urging the Government of Japan to withdraw its 2005 proposal or to use non-lethal approaches. However, Japan has issued itself permits up to the 2008–9 season. The IWC is reticent about recording the state of whale populations given the scientific uncertainty over numbers, and only issues figures for populations that have been assessed in some detail. In 2009 no estimates had been provided for minke whales or for fin whales in the Southern Ocean.[15] Fin whales are however listed as endangered by the US Fish and Wildlife Service.

Some of these whales are slaughtered in the Australian Whale Sanctuary adjacent to Antarctica. In the polar regions this Sanctuary forms part of the Southern Ocean Sanctuary. In 2008, an Australian Federal Court ruled that the Japanese whale hunt in the Australian whale sanctuary in Antarctica is illegal. It violates the Australian Environmental Protection and Biodiversity Conservation Act. The court ordered that Japan be restrained from 'killing, injuring, taking or interferring' with whales in the Australian Whale Sanctuary. The case was the first attempt to use the law to ban whaling. However, since Japan does not recognise Australian jurisdiction in the Southern Ocean the ruling was ignored. Also, the Australian Government held back from enforcing the decision. A naval vessel was sent to 'monitor' the next hunt but there was no attempt to stop it.

After the research has been conducted the whale flesh is sold to consumers in Japan under the IWC provision that carcasses be utilised, not discarded. What is occurring here is clearly commercial whaling in the guise of scientific whaling. In addition, Greenpeace has revealed that sizeable quantities of whale meat are taken off the whaling ships by crew.[16] During 2008 Japan engaged in intensive lobbying to try to get the commercial ban on whaling lifted, signalling that it will not easily give up this fight. It did not succeed. The revelation that Japan has been taking the southern bluefish tuna well in excess of its quotas and the discovery that over 2 billion Australian dollars' worth of illegally caught tuna have passed through Japanese fish markets over the past 20 years should make the international community pause when it considers whether Japan can be trusted to exploit marine resources within internationally agreed quotas.[17]

There are attempts to refashion the IWC to turn it into an instrument of conservation or preservation rather than regulation which would incorporate scientific research of a benign type. It is hopeful that these moves together with the changing sentiments within the whaling countries as mentioned above will eventually eradicate or greatly reduce the slaughter of whales.

Aboriginal subsistence whaling is also allowed under the IWC regulations. Such whaling should not seriously increase the risks of extinction and it should 'enable harvests in perpetuity appropriate to cultural and nutritional requirements'.[18] Subsistence whaling is allowed in Greenland, Siberia, St Vincent, the Grenadines and Alaska.[19] It is the responsibility of the national governments to provide the IWC with evidence of the cultural and subsistence needs of their people and the Scientific Committee provides advice on safe catch limits. Again as in the case of scientific whaling the primary decision-making power lies with the national government rather than the IWC. The whale flesh must be used only for local consumption. There is no restriction on gear used in the slaughter.

The aboriginal exemption has been very controversial. The US has been accused of inconsistency in its support of the whale moratorium while at the same time defending Alaskan hunters of the bowhead whale, a severely decimated species. The US also supported the claims of the Makah Tribe on the Pacific Coast of Washington State for aboriginal exemption in order to hunt grey whales. The Makah gave up whaling 75 years ago and have built a large recreational marina and tourist complex on their reservation. The exemption to hunt whales was granted to them in 1999.[20] This signals a broadening of the notion of subsistence whaling. Japan and Norway are also seeking exemption under the aboriginal exemption, making a mockery of the provision.

At the present time the slaughter of dolphins is not regulated by the IWC. In Japan this is under the control of the Ministry of Agriculture, Forestry and Fisheries. In 2009 this ministry issued licences to kill over 20,000 cetaceans, consisting mainly of dolphins. Some of this activity is carried out in the town of Taiji in the Kii Peninsula. The dolphins are killed in a particularly cruel manner by stabbing and spearing. They die slowly after a great deal of stress. It has been estimated than more than 400,000 dolphins have been killed in Japan by dolphin hunters over the past 20 years, despite international condemnation.[21] Numerous internet petitions to end the slaughter have been sent to the Japanese Government.

Despite Japan's push to acquire large quantities of whale flesh, the population is not responding by consuming a large amount. There is a glut of whale meat, 'Nation-wide, more than double the amount of whale meat sits in freezers than is released on to the market each year.'[22] School campaigns try to get children to eat whale and there are attempts to introduce it into fast food chains. Studies on the Taiji dolphins show that they pose a health risk for human eaters because of the levels of methyl mercury which is a neurotoxic metal.[23] If whales are not needed for food and the science based on their slaughter is questionable this increases the tragedy of their deaths. If, as seems likely from the above, whales and dolphins can feel pain then the extension of equal consideration to cetaceans would entail as a minimum requirement an end to their slaughter. Would our moral duty to cetaceans be fulfilled by simply leaving them alone? Human use of the oceans is increasing and the hazards for cetaceans escalate with this usage especially during their coastal migrations.

2. Commercial fishing

Fishing nets such as gillnets and seine nets used to catch fish are often hazardous for dolphins, porpoises, small whales or juvenile whales of the larger whale species. They become entangled and drown. The world's smallest cetacean, the vaquita, is found only in the northwest Gulf of California. It is critically endangered and is easily trapped in gillnets and as a result the vaquita are threatened with extinction. The IWC also lists the North Atlantic right whale as threatened by capture in fishing nets.[24] In 2005 it was reported that more than 300,000 cetaceans are killed every year in fisheries bycatch.[25] The problem of bycatch is particularly acute where the cetacean levels are very low. In the ocean around Sakhalin Island in the Russian Federation there are only about 100 North Pacific grey whales. These whales are endangered and any loss from fishing gear entanglement could put the survival of the population in doubt, especially as these whales also face other threats from the oil and gas operations in the area. There are methods available to avoid this bycatch such as gillnet floats that break away when hit by a whale and acoustic 'pingers' that warn marine mammals to keep clear of nets and buoy lines. The World Wildlife Fund is working globally to promote adoption of these changes and has had some success in the US.

3. Degradation of habitat

Oil spills pose a definite threat for the North Pacific grey whales. The oil and gas extraction is taking place in the feeding grounds of the whales off Sakhalin. The mothers and calves prefer the near shore feeding grounds where they are at particular risk from the commercial operations. More than 50 conservation and environmental groups endorsed a study by the World Conservation Union calling for a suspension of the oil and gas operations. In 2008 the company agreed to move the pipeline out of the feeding ground. This was a major victory for the whales.[26]

Roger Payne, from the conservation group Ocean Alliance, led a team recently concluding a five-year study of the effect of ocean contaminants on cetaceans. Industrial pesticides such as DDT and PCBs re-concentrate as they move up the marine food chain. The team found extremely high levels of these contaminants especially in beluga whales and bottlenose dolphins. The poisons inhibit a mammal's immune system, its ability to function and the development of its young. The offspring are born already contaminated.[27] Preliminary results of research currently being conducted on humpback whales off the Australian coast by Dr Susan Nash show high levels of these ocean contaminants which have been traced to their food source: krill. These hazardous chemicals in krill are not only important because they are passed on to the animals who eat them. The chemicals could lead to population declines in krill and the abolition of the whales' food source. Other chemical pollutants have been linked to a range of health problems in cetaceans including reproductive failure and cancer.[28]

Plastic debris such as cigarette lighters, toothbrushes, plastic toys, plastic bags, bottle tops and styrofoam cups commonly float around the worlds' oceans. Dolphins and whales can injest these plastics causing injury or even death.They become entangled in plastic nets or ropes abandoned by fishermen. The World Wildlife Fund estimates that around 100,000 whales, seals and dolphins are killed every year by plastics.[29]

Climate change is having an effect on some whale species. Warmer temperatures cause sea ice in the polar regions to melt and this affects the krill and other types of plankton. The krill feed on marine algae, which live in the ice.The krill larvae also depend upon sea ice for protection from predators. As the sea ice shrinks, krill populations are declining. Krill in the Southern Ocean provide the primary food source for almost all Southern Ocean baleen whales, including blue, humpback and minke whales.[30] Also as the ice retreats in warmer oceans the distance that the whales need to travel to their food source increases, possibly up to 500 kilomotres. Whales need to build up reserves in the summer feeding to maintain them for their long migrations north in the winter. The Northern Hemisphere humpback whales will encounter similar difficulties.

In the Arctic, the beluga and narwhal whales feed on cod that depend on ice-edge plankton for their survival. Decline in cod means that there may not be enough food for the whales.[31] Cetaceans make extensive use of hearing to navigate, locate food and communicate with each other. Increasing noise pollution in the oceans is interfering with their hearing abilities. Christopher Clark from Cornell University has researched oceanic noise pollution which he calls acoustic smog. According to Clark, this chronic noise from shipping is interfering with whales' well-being, limiting the range over which they can navigate, communicate and find food or mates.[32]

4. Sonar technologies

The use of sonar can be a very serious threat to cetaceans. In 2005 the Australian navy used short-range, high-frequency active sonar to detect an historic anchor. This was followed by a stranding of 130 pilot whales nearby. Mass whale stranding of three different species in North Carolina have also followed on the use of mid-frequency sonar.[33] Similar events occurred in Haro Strait off the coast of Washington State, the Canary Islands, Madeira, the Bahamas and the US Virgin Islands. In 2005 a coalition of conservation and animal welfare organisations brought a legal case against the US navy which targetted training with mid-frequency sonar used aboard US naval vessels to locate submarines and underwater objects. They argued that there is no scientific dispute about the capacity of intense sonar blasts to disturb, injure and even kill cetaceans.

> Whales exposed to high-intensity mid-frequency sonar have repeatedly been stranded and died on beaches around the world, some bleeding from the eyes and ears, with severe lesions in their organ tissue. At lower intensities sonar can interfere with the ability of marine mammals to navigate, avoid predators, find food, care for their young, and ultimately to survive.[34]

The plaintiffs claimed that sonar use violates US environmental laws, in particular the Marine Mammal Protection Act. However the Pentagon declared the navy exempt from the US law requiring steps to avoid harm to marine mammals and a Supreme Court judgment in 2008 lifted the ban on sonar testing. International bodies such as the IWC, the World Conservation Union and the UN have now taken notice of this issue and various measures are being discussed such as the designation of marine protected areas.[35] Restricting testing to areas outside those frequented by whales is the obvious easy solution.

The US navy has developed low frequency active sonar (LFA) also designed to detect submarines. This technology works by emitting intense, low frequency sounds that echo off solid objects in the ocean. These echoes are monitored by naval ships to determine whether an object is a submarine or a rock or a school of fish. The speakers used by LFA emit sounds as loud as 240 decibels. Exposure to 140 decibels causes immediate hearing damage in humans. A ship using LFA sonar can flood thousands of square miles of ocean with noise. This will cause damage to any cetaceans within the area. The level of noise emitted is many millions of times louder (the decibel scale is logarithmic) than the level known to disturb cetaceans. Some whales have been found dead on Greek beaches after LFA was used in the area. A NATO investigation into the deaths identified LFA as the likely cause. In 2000, 13 whales beached themselves during naval operations in the Bahamas. They had signs of acoustic trauma. Even noise levels as low as 130 decibels will lead whales to avoid the area, possibly keeping them away from feeding or calving grounds. 'In at least one instance, mother whales and dolphins became separated from their calves when active sonar was being used nearby, leaving the calves alone and completely defenceless.'[36] Yet with all these problems only the most cursory environmental studies have been conducted and alternative low impact technologies with similar benefits are not being developed. In 2002 some legal limits were placed on the use of this technology but it still continues in use.

5. Seismic testing

Seismic tests have been a concern of the IWC for some years, particularly those conducted by the oil and gas industries. Such tests involve the use of air guns to fire 220-decibel pulses at the sea floor, producing shock waves that can reflect undersea geology. The effects on cetaceans can be extreme, including 'permanent hearing loss, disorientation, brain haemorrhaging and death.'[37] Whales in the Gulf of Mexico and around Sakhalin are particularly vulnerable because of the amount of exploration in those areas. Sperm whales which are endangered are resident in the Gulf of Mexico and five other endangered species of whale migrate to the Gulf. The Western grey whale migrates to Sakhalin and it is critically endangered.

6. Cetaceans in captivity

It is mainly dolphins who are held in captivity. However in the 1960s the first orcas (killer whales) were captured and displayed. This practice is nearly over, with perhaps one remaining orca in captivity. The film 'Free Willy', about the attempt to release a killer whale, had quite a profound affect on public consciousness about keeping such whales in captivity. Wild orcas generally live two to three times longer than captive ones, who often die from pneumonia.[38]

The dolphin slaughter in Taiji, Japan was mentioned above. As part of that hunt an unknown number of healthy young dolphins are saved from the slaughter to be sold for a high price to the international dolphin captivity industry. These dolphins will be kept in aquariums and trained to be part of shows to entertain a human audience.[39] Although dolphins are often presented as having fun in an aquarium, the reality is different. Firstly there is the capture which involves breaking up the family group. As Cochrane and Callen state: 'Wrenching a dolphin from family members and close companions not only disrupts the social structure of a pod. Because dolphins form strong attachments to one another, it must be very distressing for the individual and those it leaves behind.'[40]

A high number of captive dolphins die within the first two years, with a possible link to captive shock. Many dolphins are kept in large concrete tanks filled with chlorinated water which stings their eyes. They require a large number of antibiotics, hormones, vitamins and fungicides as stress 'ravages their natural immunity'.[41] Attacks by one dolphin on another are not uncommon. There is limited stimulation in the tank world for these inquisitive and intelligent creatures. The tricks they are required to perform are on human demand, and food is withheld if they do not comply. These captive dolphins often fall victim to stress-related illnesses such as heart attacks and gastric ulcers, and have much shorter lives than in the wild. It is uncommon for wild dolphins to stay around human swimmers for long. Yet in a dolphin show there may be hundreds of people noisily occupying a space next to the tank. The oceanographer Jacques Cousteau had two captive dolphins for study purposes and, sadly, they hit their heads against the hard edge of the pool until they died. Their suicide led Cousteau to state: 'No aquarium, no tank in a marineland, however spacious it may be, can begin to duplicate the conditions of the sea. And no dolphin who inhabits one of those aquariums can be considered normal.'[42]

7. Whale watching

Whale watching could be viewed as a benign activity encouraging the appreciation of whales, dolphins and porpoises, and perhaps it usually is. However, in 2006 the Scientific Committee of the IWC reported that in some circumstances, whale watching and vessel traffic can have an adverse impact on some small cetacean populations. Studies in Australian waters showed that powerboats can force some dolphins to abandon their foraging for food and take evasive action. Some pods will be approached by boats many times in a day. So the harassment could have a significant effect.

In 1998 I witnessed the curious sight of a small cruiser rocking quite wildly in calm waters in Platypus Bay, eastern Australia, with the crew calling for help. Apparently they had deliberately run over the back of a humpback whale; in response the whale had positioned itself under the boat causing the rocking motion, an inventive retaliation and kind, given that one smash of the tail on the boat would have destroyed it.

PROTECTION AGENCIES

With a moral argument to support the idea that humans have a duty to promote cetaceans' well-being, and a plethora of threats, who is looking after their welfare? The Law of the Sea mentions cetaceans only briefly: 'States shall co-operate with a view to the conservation of marine mammals and in the case of cetaceans shall in particular work through the appropriate international organisation for their conservation, management and study.'[43] This is an important comment as it requires of signatories that they co-operate. For instance, they shouldn't unilaterally decide to resume commercial whaling. Second, the Law says nothing about 'utilising the resource' as it does with fisheries. It is fair to say then, that the Law promotes conservation rather than utilisation but it gives no detail on how that could be brought about.

The appropriate international agency is the IWC. Founded in 1948 as a whaler's club it changed in 1961 to a whaler's club with scientific guidance. The notion that the high seas are unowned and open for all to exploit reinforced the view of whaling as legitimate. This way of thinking, together with quotas for killing whales which were far too high, led to a serious drop in numbers. The scientific guidance came in to 'permit increases in the number of whales which may be captured without endangering these natural resources'.[44] Without accurate estimates of the sizes of whale populations this scientific guidance is always going to be limited and should operate in a precautionary manner. The IWC has now evolved to a body of 85 members containing whaling and non-whaling nations. The desire to protect the whales became dominant over the last two decades. The moratorium on commercial whaling imposed in 1986 is still in place. In 1979 the IWC established a sanctuary for whales in the Indian Ocean and another in 1994 in the Southern Ocean.

The issuing of special permits for scientific whaling and aboriginal subsistence whaling are very contentious issues, as indicated above. Also, the currently debated proposal to develop a management scheme is thought by some states to be the first step in the resumption of commercial whaling. One weakness of the IWC is that if a signatory puts in an objection to a regulation of the Commission it does not need to comply with that regulation. Ultimately then it is political persuasion rather than legal necessity that is likely to be most effective. Ethical questions about whaling have also been raised outside the IWC. Even in 1927 a League of Nations Committee of Experts report called for 'a new jurisprudence in ocean resource use' proclaiming that 'the riches of the sea and especially the immense wealth of the Antarctic region, are the *patrimony of the whole human race*'.[45] In 1946 the US acting Secretary of State referred to the responsibility of whaling nations to treat the stocks as a 'trust for mankind'.[46]

The Convention on International Trade in Endangered Species (CITES), also relevant to whales' welfare, has endorsed the moral notion of obligations to future generations and to the environment itself. This Convention has more signatories than the IWC, including 174 countries in 2009. CITES regulates and when necessary prohibits trade in endangered species and their byproducts. This works via two appendices. There is a prohibition on trade for all species listed in the Convention's Appendix I, which covers the most endangered animals and plants. Other species not necessarily threatened with extinction but ones that are at risk if trade is not closely controlled are listed in Appendix II. For these species some trade is allowed but on a restricted level and only in exceptional circumstances. All cetaceans are now listed in one appendix or the other.[47] Apart from the threats mentioned above, cetaceans have a biological nature that increases their vulnerability. They have a long gestation period and females reproduce at a late age.

As with the IWC, CITES has a policy that if signatories put in an objection to a listing of any species as banned in trade, then they can escape restrictions. Again political pressure is vital in compliance. Norway and Japan have opposed the listings for several cetacean species. This means these nations can legally trade whale products with each other.[48] Illegal trading can now be picked up through DNA marking. In a recent decision CITES rejected Japan's proposal to loosen restrictions on trade in three populations of minke whales. It was felt that such a proposal would undermine the IWC's moratorium on commercial whaling.[49] CITES acts in this way to bolster the support for the moratorium on commercial whaling.

Several non-government organisations have been working on behalf of whales. For about 30 years Greenpeace has taken up many issues to do with their welfare by direct action and use of the media. In particular Greenpeace has helped small cetaceans by destroying drift nets, releasing animals and raising public awareness about the dangerous nature of these nets. The IWC has not engaged with this issue until recently, as the death of small cetaceans such as dolphins was not seen as whaling. This is starting to change and Greenpeace's efforts have been influential. The IWC is exploring the use and effectiveness of bycatch mitigation measures, for example, acoustic devices to reduce the large numbers of small cetaceans accidentally caught in fishing gear. The IWC also now supports the UN General Assembly resolution to phase out large-scale drift net fishing.

Greenpeace helped to shut down Australia's last whaling station in 1978. Australians are now able to enjoy watching the yearly migration of humpbacks and right whales up and down the coast. For several years Greenpeace has been active in the Southern Ocean revealing the extent of Japanese hunts and the suffering of the whales. They provided graphic images despite grave personal risk. During the 2005–6 season Greenpeace zodiacs operated for 74 days, with the committed crews literally placing their bodies between the whale and the harpoon in an effort to save the whales from slaughter. They were able to slow down the hunt considerably and in one 10-day period, no whales were killed.[50] Greenpeace is also exposing the links between the Japanese company that conducts the whaling, Kyodo Senpaku, and the three large Japanese fishing companies which own the whaling company. The exposure of these links could impact on the desire by anti-whaling countries such as the US to buy the fish. So indirect pressure to stop whaling might result.[51]

The Sea Shepherd Conservation Society is a radical anti-whaling group who for several years have taken boats into Antarctic waters also at great personal risk, following the Japanese fleets and generating adverse media coverage. In 2008 two men from the Sea Shepherd boat boarded a Japanese whaling vessel to serve notice of the Australian Federal Court decision about the illegality of whaling in those waters.

The Whale and Dolphin Conservation Society supports specific scientific research on cetaceans in the wild and campaigns against commercial whaling as well as the holding of any cetaceans in captivity. Project Jonah is an important force backing the Australian Government moves in the IWC for a global whale sanctuary. The Oceania Project and Southern Ocean Whale and Ecosystem Research are both conducting valuable photo-identification studies of whales. The Humane Society International was instrumental in the legal action brought by the Australian Government against the Japanese whaling in the Australian Whale Sanctuary.

Another two large international organisations include the welfare of whales in their very broad briefs: the World Wide Fund for Nature, which has been working on conservation plans for the habitat of the grey whale around Sakhalin Island, and the International Union for Conservation of Nature, which links science to policy especially through monitoring of the state of the world's species. It provides a list of threatened species that includes cetaceans. Ocean Alliance has worked on the insidious effect of pollution on cetaceans as mentioned above. Defenders of Wildlife act as a lobby group exposing threats to cetaceans, in particular from sonar technologies and climate change.

This list does not cover all the protection agencies working for whales or all the work done by the agencies mentioned. However it is clear that the scope is enormous. It is my impression that too much of the work in looking after whales is falling on non-government organisations. The Law of the Sea passes over the responsibility to the IWC which is struggling to emerge from the whaling tradition. CITES acts as an important block to commercial whaling but it cannot tackle other issues. What the huge effort of the protection agencies illustrates, both in terms of resources and people power, is the belief that it is our duty to look after cetaceans and their habitat so that they can live and flourish in their home, the ocean.

So far the focus of this book has been on problems arising from the lack of regulation in the high seas, or inadequate controls in the coastal waters to ensure environmental protection or protection of fish and cetaceans, or a just allocation of goods. Another source of conflict arises when different cultural groups within the same area seem to have justifiable claims on sea territory, claims which have largely been ignored by the dominant group. It is to these matters that I will now turn before returning to the questions of ocean governance and possible ways forward in Chapter 9.

7
Sea Gypsies

Magtangunggu' mandelaut
Angigal mandea –
Goyak maka seloka
[Music from the sea,
dancing ashore,
waves and coconut palms]
(Traditional Bajau Laut riddle)

Pakaita tinimanta
Embal pakaita, tagu'ta-labu.
[When we use it, we throw it away;
when we don't use it, we take it out – an anchor]
(Bajau Laut saying)
C. Sather, *The Bajau Laut*[1]

maak tchi bita [Where did I put it?]
(common Moken question)
J. Ivanoff, *The Moken*[2]

A chance meeting with an Indonesian fisherman in remote waters off
northern Sulawesi in 2001 alerted me to the existence of sea gypsies.
The romance of living a life on the sea captivated me. I wanted to
find out about them, their culture, where they travelled and how
they got fresh water. If their home is the sea, how secure are they
in their home? Does the Law of the Sea offer them any protection?

SEA GYPSIES: PEOPLE WITHOUT AN ADDRESS OR 'NAMES THAT CAN BE FOUND IN BOOKS'

The term 'sea gypsy' was coined by the English explorer J. T.
Thomson in 1851.[3] Others have used the label 'sea nomads',[4] 'sea
people',[5] 'drowned people' and 'floating people'.[6] Sea gypsies live
in the waters of Southeast Asia and off the coast of Chile. I will
be focusing on the Asian sea gypsies who exist in bigger numbers

and are more widespread. These cultures bring ownership of the sea into sharp focus not as something to be exploited but as home. The sea gypsies association with sea territory is even more central to their lives than it is to the lives of coastal Aboriginal Australians – a story I will take up in Chapter 8. To make this association clear it is necessary to explore how the sea gypsies live their lives, and to understand the threats that are now facing them. It is important to see the way they are viewed by others as this helps us to understand their cultural difference. The Asian sea gypsies do not form a single cultural group. The particular sea range they inhabit has an impact on their identity. There are also historical variations or at least variations in historical accounts. Pre-twentieth-century records do not sharply differentiate sea gypsies from pirates,[7] while accounts of twentieth-century sea gypsies do.[8]

Sea-gypsy cultures can be traced back 400 years[9] and the three major groupings that exist today go back at least to the mid nineteenth century. These groupings are: (1) the Moken (sometimes called Mawken) off the coast of Myanmar (formerly Burma) in the Mergui Archipelago and extending down to waters off southwest Thailand; (2) the Orang Laut of the Riau-Lingga Archipelagos, the Bantam Archipelago and off eastern Sumatra, Singapore and southern Johor (a northern group of Orang Laut are in the sea near Phuket); (3) the Sama-Bajau (or simply Bajau or Bajaos), spread over one-and-a-quarter million square miles, from the southern Philippines, eastern Borneo and Sulawesi, south and eastward through to the waters of eastern Indonesia, to Flores and the southern Molluccas.[10]

There is very little written about the sea gypsies. The key sources are as follows: Sopher, in *Sea Nomads*,[11] presents a survey of the literature available up to 1960. Sather, an anthropologist who did field work with the Bajau in 1964, has written extensively about them, especially in *The Bajau Laut*. Jacques Ivanoff presents the reports of his father, Pierre, an ethnographer who travelled with the Moken in the 1950s and 1970s, along with his father's correspondence with a geologist, Cholmeley, who worked with the same group of sea gypsies in the 1940s.[12] Cynthia Chou, a contemporary nomadic anthropologist, spent many months living with the Orang Laut in the 1990s and continues to visit. She has been initiated into some of their religious practices. As well as revealing a close familiarity with Orang Laut practices Chou provides some valuable translations into English of sea gypsy sayings and songs.[13]

In order to understand just how independent from the land and land dwellers the sea gypsies have been we need to investigate some details about their lives. In their traditional lifestyle sea gypsies spend most of their time in small boats. A typical Moken boat called a kabang is illustrated in Figure 7.1.

A Bajau boat called a lepa is similar but with a higher bow and stern and poles protruding from both ends. Orang Laut boats are also not very different but they are high out of the water in the stern with a low bow. All the sea-gypsy boats are large enough for a family

Figure 7.1   Moken family going about daily tasks in their boat. Photo by Pierre Ivanoff, 1957. Reprinted with permission from White Lotus Press. Source: J. Ivanoff, *The Moken: Sea-Gypsies of the Andaman Sea Post-war Chronicles*, Bangkok: White Lotus Press, p. 13.

to sleep and cook in, to carry a few possessions and sometimes a dog and cat.[14] Figure 7.2 shows the sleeping and cooking area inside a Moken boat.

Figure 7.2 Inside a Moken boat. Photo by Jacques Ivanoff, 1986. Reprinted with permission from White Lotus Press. Source: J. Ivanoff, *The Moken: Sea-Gypsies of the Andaman Sea Post-war Chronicles*, Bangkok: White Lotus Press, p. 148.

Up until the last decade or so these boats were propelled by poles, oars or sails. Several boats would usually travel together and anchor together. The sea gypsies would often go ashore for fresh water that could be transported in a bamboo funnel. In calm conditions out at sea the Orang Laut knew that immersion in the sea measurably lessened thirst and there are some reports that the Orang Laut could drink brackish water without injury.[15]

The Moken trade sea produce such as fish and trepang (sea-cucumber) for rice and cloth with local land dwellers.[16] The Bajau traditionally traded fish for vegetables and fruits (especially cassava).[17] There have been some changes in the last decade, as noted later. The Orang Laut trade fish, trepang, and turtle shell for meat, eggs, rice and sago.[18]

Nations bordering the sea-gypsy waters did not know what to do with sea gypsies. In a variety of ways they were branded as pariahs. They could not be captured within the traditional social and political order because of their nomadic lifestyle. They did not have an address. It was difficult for land dwellers to keep the sea gypsies under close surveillance. When the British occupied Burma (from 1926 to 1948) they attempted to enforce rules on the Moken, but this was a 'complete failure and had to be abandoned'.[19] Accounts circulated about social customs of the sea gypsies which set them apart from land dwellers. The Malays had a belief that the Orang Laut would determine who was suitable for marriage in the following way: The man and woman would dive beneath a boat and try to catch each other. 'If they meet underwater, it means they are suited for marriage, if not it means they are not suited for one another.'[20] Yet this was not confirmed in any of Chou's numerous encounters with the Orang Laut and her informants denied that this was a practice. According to the Orang Laut and other sea gypsies suitability was determined by working together and travelling together.[21]

The poverty of the sea gypsies also set them apart from large sections of the land dwellers and formed another reason for their pariah status. The Moken were described as 'scarcely better off than the mud-fish on which they live'.[22] Despite their poverty, however, there are many accounts of their vigour and well-being.[23]

A linguistic prejudice existed too. Chou claims that the Malays did not accept that the Orang Laut spoke a language;[24] 'they did not have names that could be found in books'.[25] There are in fact a great many different sea-gypsy languages, some related to Malay and tribal languages of Borneo or Makassarese, some not. There is no evidence of an original language underlying all groups.[26] As early as 1893 Alfred Russel Wallace had translated 117 words from the Bajau language into English.[27] Sather adds to this.[28] More recently Verheijen provides a translation into English of some Bajau stories, riddles and songs.[29] Chou translates some Orang Laut sayings and poems[30] and White many Moken words.[31] Jacques Ivanoff translates Moken stories.[32]

Sea gypsies are regarded as outsiders also through the belief that they are impure, in the sense of 'a people without religion'.[33] Mythologies grew up about their origin that were thought to explain their pariah status. These usually involved a religious transgression on the part of an ancestor who behaved incorrectly or ate the wrong food.[34] Sather describes such transgression as 'an act of sacrilege, the commission of which places the Bajau Laut outside of the religiously constituted society of their sedentary neighbours'.[35]

Sea gypsies were often branded as heathen, an accusation made by Tomè Pires in the sixteenth century,[36] and a consistent theme in the literature about sea gypsies from the mid nineteenth century to the mid twentieth. Writing of the Moken in 1922, White says: 'the Mawken cannot be said to have Religion'.[37] Behind this claim is the association of religion with Christianity.[38] However, Nimmo's later study of the same people reports that they are 'still regarded as pagans by the surrounding Muslim peoples'.[39] Sather writes about the land dwellers regarding the Bajau as 'people without religion' right up to recent times.[40] Chou has uncovered a notion of purity employed by the land dwellers adjacent to Orang Laut seas such that anyone who is not a Muslim is regarded as impure, and the Islamic practices required of a Muslim include 'residing in a village with a mosque'.[41] This means that the sea gypsy, *by definition*, cannot be a Muslim and hence is impure.

These representations of the sea gypsies as irreligious must be regarded by the gypsies themselves as curious, given that they do have a religious life, with spiritual beliefs and practices. Presumably they have also puzzled over why the land dwellers do not share their spiritual beliefs. The Orang Laut for instance believe that their religious practices are the most powerful in the Malay World.[42] To highlight just a few aspects of the religious beliefs of the sea gypsies: The Orang Laut believe in the existence of sea spirits and land spirits. These spirits can be called upon for assistance, for example, to heal the sick, safeguard pregnancies and to influence souls.[43] An appeal is sometimes made to the sea spirits for control over the marine world and the winds, for instance, to bring fish towards the boats. It is understood that the effectiveness or otherwise of a fishing venture is due to the intervention of sea spirits who enter into the inner being of the fishers. The spirits are very powerful entities and great care must be taken in appealing to their forces. After the catch, food offerings are put out for them on certain rocks and the Orang Laut must bathe from head to toe 'lest the spirits devour their souls'.[44] The use of water and bathing 'functions as a boundary weakener facilitating the exit of the sea spirits from one's inner being'.[45] These spirits sometimes make an appearance and are said to be like exceptionally good-looking human beings but with bloodshot eyes.[46] These beliefs and practices are essential to the self-identity of the Orang Laut and they think the spirits are known to all (even westerners). It is just that others may not be willing to talk about their relationship with the spirits.[47]

The Moken also have a religion, a link with a spirit world, but one that differs from the Orang Laut.[48] This religion informs Moken identity. Bruno Bottignolo presents a very detailed account of the religious beliefs and practices of the Bajau that involves a complex realm of spirits differing from the religions of the Moken or Orang Laut.[49]

There are, then, systems of subjugated religious knowledge that have powerful effects on specific communities of sea gypsies and play an important role in the maintenance of their unique identities. Chou claims that the Muslim Malays in land areas bordering sea-gypsy territory, while refusing to believe that the gypsies have a religion, fear the power of the Orang Laut's contact with a spirit world. For the last few decades there has been pressure on the sea gypsies to become Muslim for this reason amongst others.[50] In some cases the land dwellers have used persuasive strategies to try to bring the sea gypsies into the Muslim worlds, to diffuse their threat, to abolish their impurity. They offer goods and houses. The sea gypsies may accept and pretend to behave in conformity with Islam, but they then often fall sick because they are living on land and so return to the sea.[51]

In a variety of ways the sea gypsies in this spatio-temporal framework are represented as an excluded group. Do these representations impact on them? There is some evidence that they do in that the gypsies appear to be timid in character, especially in their relationships with land dwellers.[52] This may, however, be part of non-sea-gypsy representation. Cholmeley, who encountered many Moken during his geological ventures, described them as 'gentle, peaceful and non-violent peoples'.[53] The three groups of sea gypsies prefer egalitarian organisation, where conflict is minimised, and leaders are chosen for experience and knowledge and are there to guide rather than direct.[54] Many of the neighbouring cultures have been strictly hierarchical and some extremely violent, such as the head-hunters of Sulawesi.[55]

There are some practical effects of these representations that the gypsies would resent, for example, the practice of throwing mud at them when they try to collect fresh water, theft of their meagre possessions, and the ransacking of gravesites.[56] However, at least until recently they probably remained ignorant of a great deal of the thinking that goes on about them given their limited contact with land dwellers except for trading. Since the sea gypsy produce is in high demand and what they want in exchange is plentiful, the negotiations can be quick. The traders are often Chinese, culturally distinct from the other land dwellers. While it might be thought that these traders could exert an influence on the sea gypsies, the traders' outsider status probably helped the sea gypsies to get around the social consequences of exchange.

The sea gypsies ignorance of local land dwellers' beliefs is borne out in the work of Cynthia Chou who, having learnt a sea-gypsy language, has been able to convey to the Orang Laut some Malay notions about the sea gypsies social mores and origin, which were greeted with disbelief. I mentioned earlier the story about marriage suitability. Also, as Chou notes, the sea gypsies do not always have a clear understanding of who is in government or of the political structure in adjacent lands. If some conflict with land dwellers arises, or they are asked for taxes or even census information, they often simply move on.[57] White presents an inadvertently amusing tale of his decade of census-taking of the Moken off Burma in the twentieth century. The sea gypsies would out-manoeuvre the census boat at sea or would pull into the shore and hide in the bushes. White writes of hopelessly wading nearly a mile 'through mud and water, walking gingerly over sharp rocks and barnacled boulders', reaching a creek with lacerated feet and finding only a few people hiding in the jungle.[58] White comments that 'a thorough census-taking under present conditions is not possible'.[59] The water skills of the Moken also allowed them to escape from the Japanese during the Second World War by swimming away.[60]

Until the last few decades the sea-gypsy cultures had proved very resilient, while over the last 500 years many land cultures have come and gone.[61] Sopher quotes Combés, a seventeenth-century writer on the sea gypsies, saying that 'These people are such enemies of the land, that it does not get from them the slightest labour or industry, nor gives the profit of any fruit.'[62] Repeatedly when the words of the sea gypsies are quoted in the literature or on tape they assert their fondness for the sea, their desire to stay there, and their belief that if they go ashore they will suffer and die. There is a belief amongst the Orang Laut that they would be struck dead by lightning if they settled ashore.[63] Sopher states that 'The Bajaos say that they get sick if they stay on land even for a couple of hours, and it is reported that if one of them should be ashore when a storm arises, he is at once taken to his boat because he feels safer there.'[64] On a recent tape recording the Moken say that they would rather die than be without their boats.[65]

THE SEA AS HOME

The sea-gypsy identity as pariah makes no sense *within* their culture. Their way of life is a choice, not something they were pushed into. They are at home in their boats and at home on the sea. Food is prepared and cooked on the boats using a hearth with earth spread around to stop the deck catching alight.[66] White mentions that on some of the Moken boats an aperture left in the bow and stern allows children to drop through the deck and crawl from one end of the boat to the other.[67]

The nuclear family in a sea-gypsy boat forms an independent unit. However, boats and families sometimes group together in a moorage or out at sea, to exchange goods and to socialise.[68] Jacques Ivanoff records sleeping on a sea-gypsy boat at sea and his memory of

> nights when the full moon makes silver reflections which sparkle on the waves, illuminating the boats, the delicate lurching, the lamentation of a woman separated from her husband and the response from another boat encapsulating the expression of one's feelings, the sharing of one's sorrow with the group, reflecting the beauty of the surroundings.[69]

The Moken are excellent sailors even in storms. It appears that no Moken lives were lost during the tsunami of 2004 even though it hit their areas. Many were near Phuket, which was devastated. Those on the sea reported being caught in a very strong current and then turning their boats to ride a small early wave as it receded into deep water, and then manoeuvring out of it. Dolphins also headed for deeper water. On the beach, gypsies working on their boats noticed unusually low water and rocks which were not normally visible. They understood what was happening and alerted others on shore to move to higher ground. A wave came with tremendous force on to the shore but didn't reach the high ground. None of the sea gypsies lost their lives though some of their boats were damaged.[70] It was a similar tale of survival in other sea-gypsy areas, despite the demise of local fisherman, for instance off the coast of Myanmar. When asked by a CBS news team why this had occurred, a Moken responded: 'They were collecting squid, they were not looking at anything. They saw nothing, they looked at nothing. They don't know how to look.'[71] As a result of the tsunami, there are now 175,000 confirmed deaths of nationals and tourists and 125,000 still missing and presumed dead.

The Moken generally inhabit around a square 100 miles of ocean. Navigation is by the sun, moon and stars and sometimes by conch shell communication with other boats. Their boats are ingenious, using simple yet effective equipment. Spearing and harpooning are the main methods employed to collect fish. The use of other technologies such as nets might have threatened their nomadic existence. Similarly, Ivanoff argues that their choice to shun agriculture is tied to a determination to retain their nomadic identity. As well as fishing, the Moken collect many shellfish and other sea creatures from inshore areas.[72]

The Bajau and the Orang Laut are also superb sailors and navigators. They live in boat villages like the Moken. Sopher writes about whole families of the Orang Laut travelling in small boats which would sometimes capsize, whereupon the whole family would swim about, right the boat, get aboard and resume their voyage.[73] He also quotes the Orang Laut as saying their boats were 'clever at playing with the waves'.[74] They use wooden anchors floating in the water to handle heavy winds. The Orang Laut fish with harpoons and also collect sea animals from the coastal regions.[75] The home range of the Orang Laut is usually about 50 miles.[76] The Bajau are more exclusively oriented towards fishing than the Moken and Orang Laut. The movements of the Bajau are determined by the fishing cycles. Communal fish drives are practised when large fish schools are attracted to the reefs.[77] The Bajau, unlike the other two groups, use nets for fishing perhaps because their range is less, usually within 25 miles in any direction from their home moorage and over tropical reefs.[78] Some Bajau sailed much greater distances in the past to collect trepang and turtle shell, and now shark fins. I will return to this below.

Many of the sea gypsies' customs reveal their closeness to the sea environment. If they observe ripples or other indications of a breeze on the water, they stop rowing, hastily raise the sail and whistle in the hope of attracting the wind.[79] Conversely they can use their knowledge of the winds to control them, to stop the winds sending ripples on the sea that could be confused with fish movements.[80] Travelling on hot days, the Moken frequently dip their heads in the sea to cool down and occasionally plunge completely into the water.[81] The Orang Laut regard sharks as brethren,[82] but the Moken take precautions with their rubbish so as not to attract sharks. White witnessed a shark attack on two Moken men.[83]

Many sea gypsy words refer to maritime phenomena much more succinctly than do their English translations; for example, the Orang Laut have a single word for 'drifting away with the sea current': *hanyut*.[84] In Bajau Laut, a communal netting in which fish are driven into a net enclosure or surround is given the label *ambit*; a lead boat in a co-operative fish drive is called an *anua'tebba'*; *binuanan daing* means sharing fish for household consumption; *dilaut* translates to 'of the sea'; *halo* means 'intermediate zone, where the sea appears green'; *magtubuk sengkol* means ritual bathing in the sea. An alliance group consisting of boat-dwelling families that sail and/or moor their boats together is called *pagmunda'*. A spirit boat used to carry away illness and misfortune is a *pamatulakan*; a navigational location-finding by the use of landmark alignments is a *pandoga'*; a *talinga* is the two ends of a semi-circle of boats assembled for a fish surround or drive; a *tebba* is a double zone of inshore waters, including both beach shallows and reef rim; and a *tunju'* is a traditional measure of length for nets (from the tip of the raised thumb to the tip of the first finger).[85]

Sea gypsies are excellent swimmers, able to swim for long distances under all manner of sea conditions, and the children get used to being in the sea early.[86] Sopher writes about the energy and enjoyment the Bajau children show in learning to paddle canoes, to swim and to throw harpoons.[87] White describes the Moken as almost amphibious.[88] The word 'Moken' signifies diving into the sea and indeed the Moken can dive 20 metres without any equipment using a feet down corkscrew motion still practised in Sulawesi.[89] Such a dive might be followed by walking on the ocean floor.[90] The Orang Laut can also remain several minutes under water and catch fish while there. The Bajau were thought to have gills because of their diving prowess.[91] Figure 7.3 gives photographic expression to a sea-gypsy child's joy and freedom in the water.

Sometimes sea gypsies ride the tides just for fun, their boats tossing violently in turbulent water,[92] perhaps akin to Australian board riders riding rips. Boat races are common too, in their own boats or with model boats.[93]

The pre-eminence of the sea is given an abstract twist in Orang Laut beliefs. According to Lenart's studies, the Orang Laut of the Riau Archipelago believe that 'the Archipelago is located at the middle of the world, which is a great disc surrounded and criss-crossed by sea, thus forming the islands. Below the ground level there is water and above there is air'.[94] They believe that this geographic centre of the world is owned by them because 'they do not live in fixed places ... Compared with them', they say, 'the Malays and others own only fixed locations, namely the villages where they live.'[95] The origin story of the Orang Laut is also very different from other such stories. They see themselves as the first people in the world. Thus

Figure 7.3   Moken children playing. Photo by Jacques Ivanoff, 1982. Reprinted with permission from White Lotus Press. Source: J. Ivanoff, *The Moken: Sea-Gypsies of the Andaman Sea Post-war Chronicles*, Bangkok: White Lotus Press, p. 119.

in the beginning ... there was already the sea, two islands and an Orang Laut couple who had a child. Because of an offence this child was cut into pieces which were thrown into the sea, and out of these pieces, islands, mountains, trees, and everything else came into being ... the first animals are thought to have descended from humans, others came later into being as a consequence of sexual intercourse with people, and in former times they spoke human language ... mountains had human characteristics, too: they were able to fight and to feud with each other.[96]

The Orang Laut believe that today there is still wood that will bleed like humans when cut.

THREATS TO SEA-GYPSY CULTURES

One of the impacts of climate change is an increase in the intensity and frequency of storms. The sea gypsies have developed clever techniques to handle stormy conditions but storms can decimate coral reefs and fish breeding grounds. This occurred around Indonesia after the 2004 tsunami. In the absence of other stresses in places such as Aceh there has been quick re-growth, but pressures abound in the sea-gypsy areas. As reefs are impacted elsewhere fishers who can move over large distances target the reefs that are still relatively healthy. The Orang Bajau in Sulawesi now compete with illegal fishers from Taiwan and the Philippines. As Asman Junaidi, a Bajau leader, says: 'With their modern equipment they take all the good fish, and just leave the small ones.'[97] In most of the waters inhabited by sea gypsies other fishers engage in destructive practices such as the use of poisons and dragnets that have severely depleted the resources of the sea. The reefs around sea-gypsy domains are also at risk of bleaching and acidification. Once the reefs go, the fish habitat goes as well. Fish are the primary food source and main exchange commodity of these communities. Higher rainfall will enhance run-off that will introduce pollutants into the sea. Corals are very sensitive to pollution. Failure of crops due to storms and droughts that are likely to result from climate change will mean that the food the sea gypsies want in addition to what they harvest from the sea may be in short supply.

Mainly as a result of competition from illegal fishing, sea-gypsy activities on land have increased and in some cases they have attempted fishing further afield. Both survival strategies are a great risk to these cultures. For instance some Moken have settled on Koh Sireh, a small island divided from main Phuket in Thailand by a street of water, but they may be moved on, as the Thai Government wants to erase their villages in order to expand the city's fishing port. The Moken have no legal right to the island. There is a plan to house them in modern apartment buildings and teach them to produce handicrafts. At present they still go to sea for seven to eight months of the year, but during their time ashore they get money from tourists who visit their villages to experience 'authentic Phuket culture'.[98]

Another Moken group have moved to semi-permanent settlements on a group of islands called Ko Surin in northern Thailand. Ko Surin is a national park and the Thai authorities are trying to enforce a fishing ban within a three mile zone off these islands. This group are fairly safe from tourist development but face restrictions in the areas where they can fish, collect sea creatures and chop down trees to re-build boats, a vital part of their culture. Thus the attitude of Thailand seems to be limited tolerance of the people but lack of respect for their culture. The Moken near Myanmar are threatened by new military bases on islands near the waters where they live.

The Orang Laut in Malaysian waters face a different sort of co-option. As mentioned they have been offered incentives to come ashore and convert to Islam. Chou claims that 'The Malays generally accept the Orang Laut as indigenous people of the region' but they are not part of the ordered Malay world because of their refusal to give up a nomadic lifestyle and their refusal to accept Islam. The Malays grudgingly accept the powers of the Orang Laut spirits but this is not to promote the integrity of the Orang Laut culture, more to diminish its supposed power.[99]

SEA BORDERS, SHARK FISHING AND CULTURAL SURVIVAL

The Bajau are still the most robust of the sea-gypsy cultures. Although they are found in many Southeast Asian waters and prefer to stay within about 25 miles from their moorages, the pressures of overfishing and destructive fishing practices by other fishers have now forced many to attempt to move further away. For centuries some had ventured south to Ashmore Reef and even as far as the Rowley Shoals, west of Broome in Australia. They had names for these islands and refer to them in poetry and song as their 'gardens in the ocean'.[100] Traditionally they came to these areas to collect trepang and trochus shells. Now these voyages are increasing and more commonly the aim is to collect shark fins.

The sea gypsies face competition from other Indonesians for these shark fins, fishers who use much more efficient and sophisticated technologies. The sea gypsies use traditional methods involving a 'shark rattle', 'a bamboo pole, at the bottom of which are a number of halved coconut shells stacked on top of each other'.[101] This is lowered into the water from the boat and rattled. The noise attracts the sharks that are then caught on baits from hand lines. The shark is hauled alongside the boat and the fin cut off and dried on the boat. The fins are taken back to Roti in Indonesia and sold to traders from China and Taiwan for shark-fin soup. This is hard and dangerous work in the traditional style and the competing fishermen may well kill off all the sharks soon anyway.[102] Another consideration is the appalling waste of the lives of these sharks to provide a delicacy in the gourmet food market, and the cruelty of the practice as the sharks, finless, are left to starve.

As these voyages are long, women and children are usually left behind and some have settled on the edge of a village in Roti,

Indonesia. In 1994 Australia declared a 200 nautical mile EEZ that incorporated many of the fishing areas of the sea gypsies. Indonesia followed suit, but because of the complex nature of archipelago sea boundaries, in this case the starting point for the zone is Sumba. The result has been that Roti is much closer to many 'Australian' islands than they are to Australia. For instance, Ashmore Reef is 78 nautical miles from Roti and 190 from the Australian mainland.

In a Memorandum of Understanding signed in 1974 by Australia and Indonesia, traditional Indonesian fishermen were given access to Ashmore Reef and four other small reefs to fish and collect trepang and trochus.[103] Ashmore Reef was later declared an Australian Nature Reserve, so all fishing and collecting was banned and still is.[104] If the fishermen are found there or away from the other four small reefs that are now becoming fished out, or if they are deemed not to be traditional fishermen, then they can be regarded as illegal. They may be apprehended by Australian naval patrols, towed to Darwin or Broome and tried for illegal fishing.

The definition of 'non-traditional' is loose but may involve simply having a small engine, useful for handling bad weather. The court process is slow and the men are imprisoned if they are found guilty as the fines imposed cannot be paid. The boats are burnt even if the fishers are innocent as they are deemed unseaworthy by Australian courts. The fishermen regard the burning of their boats as a type of murder, and when a boat is burnt they experience this as a death.[105] The whole community is affected. Each boat may have seven to nine crew who have families to support. The loss of a boat 'seriously jeopardises the livelihood of at least 30 or 40 people'.[106] Balint documents the plight of some of these families left without support. She writes about one fisher's grandmother who walked 16 kilometres a day to beg for rice.[107] Le Bau, an old Bajau man, expresses his frustration at the court process when addressing the magistrate as follows: 'It would not matter what I said to you, you will not listen to me and you will not understand.'[108]

There is a question about the fairness of drawing maritime borders in the way indicated given the prior occupation of sea peoples. They are made invisible, just as the Aboriginal presence on mainland Australia did not count for ownership claims. Putting that problem to one side, it is difficult for the fishermen to stay exactly within the specified areas given their lack of sophisticated equipment to both detect and control where they are. One detained Bajau man said: 'As far as the border between Australia and Indonesia is concerned, I can't draw it. Because the sea has no borders.'[109]

As nation states pursue border politics not only on land but also in the sea, the viability of these sea-gypsy cultures is looking questionable. Grudging short-term allowances relating to land and sea territory have enabled the survival of a few thousand sea gypsies up until the present day, but the future is uncertain. The Law of the Sea is definitely part of the problem in that it has allowed nation states to effectively take over and control sea territory up to 200 nautical miles from the coast. However there is also the issue of small, traditional fishers being eased out of any territory, whether coastal or in distant waters, by bigger, more efficient, more ruthless operators. For the sea gypsies this is particularly tragic as it amounts to denying them a home.

Some other indigenous groups live close to the sea and are increasingly asserting their rights to sea territory. These claims are morally strong but how can they be established in courts of law?

8
Indigenous Sea Claims

...the waves when they go in and when they go out. It is all related to us – we are from that place.

Dhuwarrwarr Marika, Saltwater[1]

OWNERSHIP AS BELONGING

Unconstrained by western philosophical ideas about land ownership and untouched by the development of cities, indigenous peoples before conquest enjoyed a very close association with their territories and the animal life therein. This is evidenced in art, oral histories and contemporary writings. The conceptual links varied in different places but one common theme is the deep sense of belonging to territory, a notion quite different from land ownership in the western tradition based as it is initially on occupation and later on the acquisition of a legal title to a parcel of land upon payment. Ownership thus conceived appears quite superficial in comparison to indigenous peoples' sense of belonging. That sense does not seem to depend on living in a confined 'private' space, since it was felt as keenly by the Aboriginal peoples of the Western Desert in Central Australia, whose range extended over hundreds of kilometres, as by the inhabitants of Croker Island, a small space in northern Australia.[2]

Another trait that indigenous peoples share is a lack of a clear distinction between belonging to the land and belonging to the sea. The Maori writers Nin Tomas and Kerensa Johnston express this as follows:

Coastal tangata whenua [People of the Land] have always asserted their 'rangatiratanga' [Maori stewardship and ownership] ... over the coast and surrounding seas [of Aotearoa: New Zealand]. ... areas of the sea have been jealously guarded mai raano [since time immemorial], with pou [sign-posts] being erected and rahui [restrictions] being set up to notify group territoriality. In pre-European times wars were fought between rival groups to protect rights to the sea and the foreshore and to oust interlopers. Taniwha [spiritual sea creatures] often acted as guardians of those rights. Knowledge of their presence throughout the area and their association with specific human whakapapa (ancestral lines), identified rights to the area as being vested in particular groups. These particular tikanga [values] were an accepted part of Maori custom law.[3]

Ownership of the sea in this perspective relates to kinship links more than to occupation and survey definitions.

The people of the Haida Nation consider themselves the rightful owners of Haida Gwaii, the Queen Charlotte Islands of British Columbia, Canada. In Haida these are also called Xhaaidlagha Gwaayaai: islands on the edge of the world. The Constitution of the Haida Nation says 'Our culture is born of respect; and intimacy with the land and sea and the air around us.'[4]

Meyers et al., writing in the journal *The Australian Institute of Aboriginal and Torres Strait Islander Studies*, describe the view of the Australian indigenous population as follows:

Ever since Aboriginal people have occupied Australia, they have used the sea and its resources. Where there are useful materials and food to be taken from the sea, they are taken. Aboriginal people have always travelled across the sea and traded goods obtained from the sea across vast distances of the interior. They do not recognise any *de jure* distinction between land and sea. Aboriginal estates that border the sea normally extend into the sea with no artificial distinction created, as in Western law, between the aquatic and the terrestrial. Aboriginal religious Law recognises significant places in and above the sea in the same way these places are recognised on land.[5]

These ideas are expressed less succinctly but more poetically in the essay 'We Always Look North'.[6]

With this broad idea about territory and a rich sense of ownership as belonging, it is easy to imagine the disbelief, confusion and outrage of the indigenous peoples when western conquerors came and asserted land ownership. As outsiders these conquerors could not experience a sense of belonging to the land or sea. When the Maori chiefs were asked to comment on the proposals for ceding the sovereignty of their country to Queen Victoria, Chief Te Kemar accused the British of taking their land and demanded that it be returned. Addressing the Governor, he said in Maori:

> O Governor! Go back. Let the Governor return to his own country. Let my lands be returned to me which have been taken by the missionaries – by David and by Clarke, and by who and who besides. I have no lands now – only a name, only a name. Foreigners come; they know Mr. Rewa, but this is all I have left – a name. What do Native men want of a Governor? We are not whites, not foreigners. This country is ours, but the land is gone. Nevertheless we are the Governor – we, the chiefs of this our fathers' land. I will not say 'Yes' to the Governor's remaining. No, no, no; return. What! This land to become like Port Jackson and all other lands seen [or found] by the English. No, no. Return. I, Rewa, say to thee, O Governor! Go back.[7]

Despite these strong words, shortly after this speech in 1840, Chief Te Kemar along with the other Maori chiefs signed the Treaty of Waitangi with the British turning Aotearoa into the British colony of New Zealand, but securing some territories for the Maori.[8] There was clearly confusion about the extent of these territories, with at least some of the Maori chiefs believing that all the land would be regarded as theirs since that is how they viewed it. However in the following years it became clear that the Maori were to be granted their villages, pa [buildings], burial grounds and cultivated lands only. Land beyond these areas was considered 'wasteland', the Crown took ownership of it and sold it to settlers who by 1861 outnumbered the Maori.[9] The Treaty and the 'clarification' did not put an end to disputes over sovereignty and territory.

In Australia it was not just the areas beyond indigenous settlements that were regarded as wastelands, but the whole land. This was encapsulated in the idea of *terra nullius*. There was no recognition of indigenous ownership of land. It is debatable whether there was recognition of Aborigines as people. Alfred Searcy, a customs officer in northern Australia from 1882 to 1896 writes about the sale of Entrance Island where Aboriginal people were living: 'Of course the niggers would give trouble but settlement by determined men has a soothing effect ... wholesale murder is inevitable in developing a new country.'[10] This idea that the land belonged to no one before the arrival of the British was overturned in the 1992 Mabo judgment in the Australian High Court, initially a strong endorsement of land rights for Aboriginal peoples, though it became watered down by later government decisions. Land claims are still being contested.

The notion of *Mare nullius* is similar to the idea of the freedom of the seas. The sense of ownership of land encapsulated in western thought involves the idea of partitioning a segment of space and marking this out, if not actually on the space with perhaps a fence then at least theoretically in maps and measurements. This is not easily translated into ownership of the sea, as I discussed in Chapter 1. However, the idea of ownership as belonging can easily extend to sea territory especially when no radical divide between the two realms is accepted. This clash of frameworks for understanding ownership has been one reason why it is much harder for indigenous peoples to assert their rights over the sea than over the land. In addition, with land claims, conflict arises with private owners or the government, but with sea claims, conflict arises with the government and more importantly with international law. The idea that anyone can navigate and fish the oceans, strongly defended by Grotius as we saw in Chapter 1, still exists in international law subject only to the constraints of some fishing agreements. Around many coasts there is the legal possibility of exploiting natural resources such as oil and gas in the sea, in territories that indigenous peoples think belong to them. Such exploitation may be opposed by the indigenous groups but with their claims to ownership of the sea having no legal standing such opposition cannot get very far.

CONTEMPORARY ATTEMPTS TO ASSERT OWNERSHIP OF THE OCEANS BY INDIGENOUS GROUPS IN NORTH AMERICA AND NEW ZEALAND

Despite the conceptual and political difficulties, in the last decade or so several indigenous groups living in different continents have mounted legal cases to try to gain ownership of sea territory. I will briefly mention some of these initiatives before examining in greater detail the moves made by indigenous Australians to gain ownership of the sea, which had a positive outcome in 2008.

During 1998 five Alaskan native villages took the Secretary of Commerce to court, asserting aboriginal title to portions of the US outer continental shelf in Prince William Sound in the Gulf of Alaska, and the lower Cook Inlet regions of Alaska. In particular they argued for an exclusive right to fish. Their claim was based on exclusive rights of occupancy and use of the continental shelf over a 7,000-year period, and on the fact that they still depended on fishing for subsistence. The court concluded that they did not have this exclusive right and thus denied them title to the waters claimed. The right of navigation, the public right to fish and national defence were used against the claimants.[11]

The Haida nation of British Columbia state in their constitution that their 'territories include the entire lands of Haida Gwaii, surrounding waters, sub-surface and air space recognising the independent jurisdiction of the Kaiganii. The waters include the entire Dixon Entrance, half of the Hecate Straits, halfway to Vancouver Island and Westward into the abyssal ocean depths.'[12] This statement is at odds with the assertion of ownership over the offshore areas by the Province of British Columbia.

In 2002 the Haida Nation began legal action against the Government of British Columbia seeking recognition of sovereignty over their lands and waters.[13] The outcome was that the Crown must consult with the Haida Nation and 'where indicated accommodate aboriginal interests' but that phrase was left vague. The Haida are particularly concerned to block attempts to mine oil and gas offshore if the current moratorium is lifted. Speaking for the Haida, Guujaaw said: 'We don't believe offshore oil and gas can be safely obtained, the technology doesn't exist, and we are not prepared to see offshore oil and gas drilling in any waters within a 200-mile limit surrounding Haida Gwaii.'[14] By 2009 the moratorium had not been lifted but Shell and Chevron still have oil and gas tenure rights in the disputed region. Both companies are contesting a 110-wind turbine proposal for the offshore areas of Haida Gwaii which has the agreement and partial involvement of the Haida.[15]

In New Zealand, the Maori were granted fishing rights in the Treaty of Waitangi, but from the 1860s onwards restrictions were imposed on these rights. These restrictions accelerated in the twentieth century. The Treaty gave sovereignty to the British with certain protections for the Maori but then the British used their sovereignty to undermine these protections. There have been battles over ownership of the sea and fishing rights. One recent case looked at the ownership of the foreshore and seabed of Marlborough Sounds. This modest claim was made 50 years after an attempt to claim ownership of the oceans around New Zealand. In the *Tai Tokerau* case heard by the Maori Land Court in 1955 the Maori spokesperson argued that

The water surrounding New Zealand should be held in trust for all Maori, because their ancient canoes had crossed and re-crossed the Pacific Ocean long before Europeans discovered Moana (the ocean). The kaumatua [Speaking Elder] stated that they had a duty to their ancient tupuna [ancestors] to ask the Court to recognise their interests in the ocean, as a mark of respect to Moana's wisdom ... in making this part of the world so extensive that New Zealand could be fished from the sea far away from lands involved in troublesome conditions.[16]

This case was dismissed as the court decided that it did not have jurisdiction.

In 2003 the Maori Land Court was given jurisdiction to hear the *Marlborough* case. This resulted in public outrage even though there was no certainty that the court would find in favour of Maori claims. One ramification that seems to have been of concern was the possibility of having to recognise territorial rights to marine areas that are not covered in English law. The government made clear its intention to legislate to protect the foreshore and seabed for all New Zealanders. The Foreshore and Seabed Act passed in 2004 secured Crown ownership of these areas. Maori are entitled to take claims concerning territorial rights to the High Court but their rights and remedies are of a lesser kind than would have been possible previously.[17] The United Nations Committee on the Elimination of Racial Discrimination released a report on this legislation in 2005 saying that it 'appears ... on balance to contain discriminatory aspects against the Maori'.[18]

AUSTRALIAN HIGH COURT DECISIONS ON SEA RIGHTS

Claims about ocean ownership and occupation have been tested through the Australian courts for some years. Grotius' reasoning has been employed. In particular his arguments for the freedom of the seas have been used by Australian courts to deny Aboriginal sea rights. The first case to reach the High Court of Australia was brought by Mary Yarmirr and others on behalf of several Aboriginal groups from Croker Island off the Northern Territory in 2001.[19] My aim in discussing this case and the following Blue Mud Bay case is not to fully examine arguments for and against indigenous sea rights but rather to show how current legal reasoning is still caught up in Grotius' argument from occupation and his defence of the public rights to fish and to navigate.

The claimants first argued that the native title rights and interests they hold confer possession, occupation, use and enjoyment of the sea and seabed within the claimed area to the exclusion of all others.

The claimed area was 'as far as the eye can see' from Croker Island. The Federal Court judge acknowledged that they had a full range of native title rights to fish, travel over the sea and to safeguard spiritual sites but he did not grant these rights exclusively.[20]

The Commonwealth appealed to the High Court claiming that native title rights could not extend into the sea. Mary Yarmirr appealed on the grounds that the Native Title Act should allow Aboriginal people exclusive rights over the sea country just as they enjoyed exclusive rights over land country on the island. The majority judgment of the High Court delivered in 2001–2 dismissed the Commonwealth's appeal. The judges accepted that native title rights and interests do exist over the claimed seas but not in an exclusive sense.[21]

The claimants wanted to be able to exclude others. Their traditional beliefs entail exclusive rights, their past actions, for instance in dealing with the Macassans over centuries, showed that they had exercised this right.[22] The Macassans came to Croker Island and surrounding areas to gather trepang, negotiating this activity with Aboriginal people. The judges said that the claim to exclusion was inconsistent with the public right to fish, and the rights of innocent passage and navigation, that is, with the rights that Grotius had defended and which are laid down in the Law of the Sea. The claimants were persuaded that they couldn't win a challenge to international law so they put in a revised claim – that their rights and interests be acknowledged subject to a qualification recognising the international rights. This was less than they had wanted and felt entitled to. In particular the claimants have a current interest in preserving sacred sea areas against tourist traffic and in protecting in the inshore areas the fish stocks that form a major part of their diet. The High Court decided that the claimants have a right to visit and protect places within the claimed area that have cultural and spiritual importance, but if there isn't control of access to and use of such areas how can they exercise protection? Similarly the claimants have a right to fish these waters, but if there is no exclusive right how can they protect their food sources from large-scale non-Aboriginal fishing ventures?

Even the High Court Justice Kirby, who clearly wanted a result favouring the claimants' view, found in his minority judgment that he could not question the rights of Grotius. He stated that the sea by its nature is not capable of being occupied, as non-indigenous perspectives understand that term, but 'the claimants argument that their traditional laws and customs do recognise a form of "occupation" and "possession" of the waters has obvious factual merit'.[23] As a result he wanted to give the claimants the power of effective consultation and a power of veto over other fishing, tourism, resource exploration and like activities within their sea country 'because it is theirs and is now protected by Australian law'.[24] It is hard to see how this position avoids inconsistency. If Justice Kirby accepts that the sea in question cannot be occupied (in the English sense) then he believes that we have to infer that it cannot be owned and that there is, for instance, a public right to fish. At the same time, he wants to accord Aboriginal people rights of veto over non-Aboriginal fishing based on a different, Aboriginal, type of occupancy. Justice Kirby was trying to find a way around the international law without putting out a challenge to Grotius' list of rights.

Another complexity of this case is that Australia is supposed to have sovereignty in the territorial sea. If occupation is important for exclusive rights then it would seem to be important for sovereignty too. Yet Justice Kirby states that the sea by its nature is not capable of being occupied and this is his reason for denying exclusive sea rights to indigenous peoples. To be consistent he should deny that Australia has sovereignty over the territorial sea, but he has not done so. I think that to hinge sea rights or sea sovereignty on a notion of occupancy is always going to be difficult as no clear meaning can be given to the notion of the sea being occupied. We can talk about sea borders with GPS measurements but that is a different idea from occupancy.

A more recent case regarding indigenous sea rights reveals further dimensions of the tangles in reasoning stemming from the view that ownership arises from occupation. This is the case bought by Gumana for several Aboriginal groups against the Northern Territory Government, known as the Blue Mud Bay case. This native title claim seeks the right to exclude others from the intertidal zone and from the sea around certain sites of significance. The territory in question is part of Blue Mud Bay in east Arnhem Land, Northern Territory. The intertidal zone may sound like a small area. It is the difference between the high and low water marks. In Blue Mud Bay the tidal difference is generally about 2 metres. Even a 2 metre tidal difference can amount to a big change in sea space if the sand gradient is very small, and in northern areas of Northern Territory tidal differences of 7 metres are common. In addition, if this case succeeds, traditional owners will be given exclusive access rights to 80 per cent of the Territory's coastline and river systems.

In 1980 the Arnhem Land Aboriginal Trust was granted ownership of the land area of Blue Mud Bay. The seaward boundary of the relevant land was defined as the low water mark. One question asked by the Federal Court Judge Selway was whether this grant conferred on the Land Trust the exclusive right of occupation over the whole area of the grant. In particular he wanted to ascertain whether it excluded any public right to fish or navigate. Judge Selway stated that given that there was a reserve over the area 'then the Aboriginal peoples did occupy the intertidal zone in a similar way to the land. Hence the public right to fish and to navigate has been abrogated in this area.' However, he felt that he couldn't conclude this as he was bound by the Croker Island case that upheld the public rights even though he felt uncomfortable with it.[25] Judge Selway died before final orders were made and Judge Mansfield took over. He denied an exclusive right to ownership of the relevant waters and so found in favour of a public right to fish and navigate. However he gave the Aboriginal claimants the right to control access to these areas, mirroring Justice Kirby's apparent inconsistency. He did not grant the Aboriginal claimants the right to control access to certain sites beyond this zone that they wanted to maintain and protect. However he recognised that there is a right within Aboriginal beliefs to control access of other Aboriginal people who agree to be governed by traditional laws and customs.[26] This is weak, as it does not prevent others from, for instance, driving speedboats to sacred sites in the ocean such as fragile rocky outcrops.

The Appeals Court didn't alter the Federal Court's finding about sacred sites. However in a surprise decision they altered the finding about inshore areas. The judges considered the Croker Island case and decided it was 'plainly wrong' and 'ought not to be followed'. They argued that given there was a land grant to the low water mark this conferred an exclusive right over the intertidal zone of the land grant. There is no public right to fish or navigate in this area. Fishing in the water flowing over the intertidal zone of the land grant from a boat would be 'no less a trespass ... than would fishing from the surface of the land in that zone'.[27]

The Appeals Court didn't take up Selway's claim that the indigenous people occupy the waters of the intertidal zone. It's the fact that they occupy the land, which is regularly underwater, which is the basis of their exclusive rights to the water above it. This is still not a challenge to Grotius as occupation is the pivotal point in assigning rights. It is because indigenous peoples cannot be said to occupy the waters around their sacred sites beyond the intertidal zone that they were not granted exclusive rights of possession to them.

In one of the most important High Court cases to be conducted in Australia, the Appeals Court finding was upheld.[28] The majority decision of the judges was that there is no public right to fish in the intertidal zone of the contested waters. They claimed that this right had been taken away by the state Fisheries Act which controls where fishing can take place in the Northern Territory. Under this Act, fishing licences are distributed for certain areas. So the High Court judges found that the Federal Court Judge Mansfield was wrong to assume that a public right to fish ruled out Aboriginal claims to exclusive fishing rights. Such a finding up to this point could be viewed as the first major rejection of Grotius' views about the public right to fish laid down by an Australian court. Prior to this decision it would have been reasonable to assume that the public right was primary even if it could undergo modification with the requirement for licences or the need to abide by quotas. Now the High Court has decided that the public right to fish does not exist in Northern Territory waters. The Fisheries Act regulates fishing. That is what is primary.

The High Court also addressed the question whether there was any competition between the rights under the Fisheries Act and the grants given to the local Aboriginal people under the Land Rights Act. Here they decided that the Land Rights Act should take priority. The majority decision stated 'The Land Rights Act ... must be understood as granting rights of ownership that for almost all practical purposes [are] the equivalent of full ownership of what is granted ... it is a grant of rights that include the right to exclude others from entering in the area identified in the grant.'[29] The High Court decided that the grants related to the land to the low water mark and that entry within the boundaries of that area, whether covered by tidal waters or not, was prohibited under the Land Rights Act, unless the owners gave permission.

From here the reasoning becomes convoluted. It would be easy to assume that if Aboriginal people are granted ownership of an area covered by water (at least on a regular basis) then they are granted ownership of this bit of sea, but not so. The judges want to define this part of the sea as land: '"Aboriginal land" when used in the Land Rights Act should be understood as extending to so much of the fluid (water or atmosphere) as may be above the land surface within the boundaries of the grant and is ordinarily capable of use by an owner of land.'[30]

Hence the Aboriginal claimants as owners of the land to the low water mark are owners of this 'land' even if it looks like sea. They have exclusive fishing rights and there is no public right of navigation in this area. It would take an Act of Parliament to change this outcome. High Court Justice Kirby said that if a government wishes 'to qualify, diminish or abolish these legal interests it must assume electoral and historical accountability'.[31]

The ramifications of this decision will be felt at the state and national levels, possibly also internationally. There is no difference between the land grants in Blue Mud Bay and most of the Northern Territory coastline. Where the tidal differences are large, the legal title to huge tracts of sea will be justified under the claim that the intertidal zone is Aboriginal land. As the Aboriginal owners have the right to exclude others from this land/sea they will be able to make use of the territory for their economic benefit either by direct exploitation of the marine life or by the selling of licences to others to exploit the fish, turtle, trepang and other creatures. This will enable these Aboriginal groups to move out of poverty and help mitigate the sense of injustice, felt for 200 years, of white control of the seas around their lands.

The national impact will flow through to other Aboriginal communities that have land title down to the sea. However with different state laws this may not work out in an identical fashion in each state. The northern communities have the most to gain as it is here that the intertidal differences are greatest.

The other national impact concerns the Croker Island decision that held up sea rights claims for seven years. This decision is no longer regarded as the benchmark. It defended the public rights of fishing, innocent passage and navigation against the exclusive use of Aboriginal peoples. Given the legal criticisms now made of this judgment the possibility arises that there might be grounds to agree to exclusive Aboriginal use of the sea or even ownership of the sea. This is what the High Court in the Blue Mud Bay case decided, but their grounds were odd. They felt that they needed to declare certain sea territories to be land in order to award ownership of the sea. However this could be seen as a type of legal conceit. It may be that the judges were motivated to award sea rights to Aboriginal owners of the adjoining land for other reasons. Perhaps they accepted the argument that the owners do not see any significant distinction between land and seas; that for centuries their ancestors had occupied and used these waters, that Aboriginal people felt a spiritual link with this sea territory, that granting these areas to

Aboriginal people for their exclusive use would help redress the injustice of taking their territory away from them and so on. Though these thoughts do not surface in the written judgment they are likely to be behind the strange legal reasoning that land can equal sea, *otherwise why take that path?* If Australian legal thinking has these sorts of motivations then it will not be a big step to say that sea territory which has long historical links with aboriginal use and spiritual beliefs should be open to ownership claims. As a matter of justice, it would seem that these claims should be considered quite independently from anything to do with intertidal areas.

The High Court decision was made shortly after the recently elected prime minister Kevin Rudd made a formal apology to Aboriginal people for past wrongs, including injustice in the law. Although the courts are supposed to be independent of government it is difficult to imagine that this apology, regarded as very significant by Aboriginal and non-Aboriginal Australians, has no guiding force. If it does then the suggestion would be to find a way to give some sea territory back to Aboriginal people for the general reasons just listed, and it's here that an international impact might ensue. We saw above that many contemporary struggles over ownership of the sea by indigenous people share common elements. If one nation is able to cede some sea areas to indigenous ownership it could serve as a precedent for other battles. Despite what the High Court judges said in the Blue Mud Bay case, if we think beyond the legal conceit, their decision does challenge the public right to fish and to navigate and it does acknowledge the claims of Aboriginal people to ownership of the sea. Once these aspects of international law are thrown into question, there should be an opening for other indigenous groups to claim legal title over the sea.

INDIGENOUS SEA RIGHTS AND ENVIRONMENTAL THREATS

It would be a sad irony if just at the time some indigenous peoples are accorded sea rights the sea realm is impacted by climate change and seriously degraded, affecting sustenance and livelihoods. The coastal indigenous communities of northern Australia are very dependent on the marine harvests from coral reefs, under threat from warming oceans, and wilder seas produced by storms and acidification. Other fish breeding grounds, for instance around mangroves, may suffer from the likely increased storminess.

Indigenous groups in the Arctic are already experiencing the impacts of climate change, with the Shishmaref community in

Alaska facing relocation and the melting sea ice in Greenland posing problems for hunters. The changing conditions are already depleting the prey animals: seals, polar bears and Arctic fish are under stress in various ways, as discussed in Chapter 2. These groups also fear activities that may be aided by climate change, such as oil and gas exploration and exploitation and the opening of the northwest passage. The former are of concern because of the increased likelihood of oil spills or gas pollution. Opening the passage would increase shipping and pollution in the sea. Sheila Watt-Cloutier, former head of the Inuit Circumpolar Conference representing 150,000 people in Alaska, Canada, Greenland and Russia has tried to put pressure on the major emitters of greenhouse gases to cut emissions from burning fossil fuels, saying 'climate change is a form of human rights abuse'.[32]

9
Protection of the Oceans

The sea can form the property of a moral person who would be the international society of states.

Geouffre de la Pradelle, 'Le droit del'État sur la mer territoriale'[1]

OWNERSHIP OF COASTAL AREAS

Grotius thought that private or national ownership of maritime coastal areas was acceptable but that the great expanse of the oceans should be free. Even in the coastal areas he thought that there should be a public right to fish and to navigate. In the twenty-first century states can assert sovereignty over 12 nautical miles from their coast retaining these public rights. Although some of the problems raised in this book could be sorted out if this territory was given over to international ownership other problems would be generated. Coastal states would in particular have concerns about security.

However, following on from the last two chapters, there are good reasons why national sovereignty over the territorial sea should be relinquished in some areas where the viability of an indigenous culture now depends on having some ownership of sea territory. This concerns the sea gypsies of Southeast Asia who are desperately trying to hold onto their way of life and also some other indigenous groups which have ownership of land bordering the sea or justifiable claims on that land not yet recognised in law. The support for these assertions is partly a moral one and partly one of expediency. It is increasingly recognised that promoting cultural diversity is a global public good, benefiting not just the local indigenous groups for instance but enriching humankind as a whole. The role of biodiversity in maintaining human food security and health through medicines and environmental conditions conducive to human flourishing is also acknowledged. Indigenous communities are thought to play an important part in preserving biodiversity.[2]

The Convention on Biological Diversity that came into force in 1993 states that

> Subject to its national legislation, [each Party is obliged to] respect, preserve and maintain knowledge, innovations and practices of indigenous peoples and local communities embodying traditional lifestyles relevant for the conservation and sustainable use of biological diversity and promote their wider application with the approval and involvement of the holders of such knowledge, innovations and practices and encourage the equitable sharing of the benefits arising from the utilisation of such knowledge, innovations and practices.[3]

In 2009, the Convention had been ratified by 178 states but not by the US. Despite the qualifications there is a risk that biological diversity will simply be treated as a resource and attempts could be made to commodify traditional knowledges. The 'Native Cultures and the Maritime Heritage Program' runs the risk of doing just that. This Program organised by the National Ocean Service in the US aims to 'support research into seafaring traditions and the preservation of maritime folklore and knowledge'.[4] The Program also aims to recognise the special cultural ties certain areas have to indigenous seafaring groups. However there is no suggestion of relinquishing ownership of these areas. Marine sanctuaries have been set up on the Olympic Coast, Channel Islands and in Hawaii but the indigenous people with links to these areas are accorded only a consultative role, 'assisting the sanctuary to shape policy' with little possibility of significant cultural revival. Preserving something of the heritage is not the same as allowing the culture to exist as a living reality.

It is clear from the discussions in Chapters 7 and 8 that it is the lifestyle of indigenous communities that helps to protect biodiversity. It is important then to maintain those lifestyles in order to reap the benefit of biodiversity conservation. The way to further this aim is to give indigenous communities ownership of land and sea territory. According to Langton, Rhea and Palmer, the Convention holds on to the idea of national sovereignty and 'does not recognise sub-rights or *a priori* rights against the nation state'.[5] However Langton et al. acknowledge that some nations have 'enacted legislation that recognises the unique position of indigenous peoples and local communities *vis-à-vis* the nation state'.[6] The Blue Mud Bay decision, though coming four years after their comment, is a good illustration.

Threats to indigenous cultures can arise from western conceptions of conservation especially those that aim for protection of critically biodiverse areas where these communities are living. One group of sea gypsies, the Moken in Southern Thailand, were excluded from participation in the management practices in the areas where they lived when these land/sea territories were declared 'protected areas'.[7] This practice can have the result of breaking up the cultural groups and even undermining conservation efforts. Langton et al. do, however, see a change occurring where governments recognise the value of maintaining traditional knowledges and ways of life as an important direction in biodiversity conservation. They also point to the fact that the International Union for the Conservation of Nature (IUCN) recognises 'the rights and interests of indigenous people to own, manage and sustainably use areas of land and sea of high conservation value'.[8]

It is not only the national sovereignty over the seas near the coast that needs to be questioned where there are indigenous claims; Grotuis' two public rights also need to be scrutinised. The public right to fish can destroy indigenous food sources, for example by the use of cyanide to capture fish in Southeast Asia and the bottom-trawling fishing techniques used by large-scale commercial trawlers off northern Australia. The public right to navigate can lead to disrespectful and harmful practices around sacred places in the sea in particular near special rocks or islands. As recognised in the Blue Mud Bay case, a stricter type of ownership needs to be accorded to certain sea areas than has been the case within the territorial sea. Some areas need to be regarded as off-limits to the public unless permission and/or licences are granted to enter. This could go some way towards reversing the unsustainable fishing and collecting practices that are currently dominated by non-indigenous commercial interests together with the maritime recreational activities that do not rate the continuing existence of these cultures highly. Money from licences could greatly assist in the ongoing viability of these cultural groups.

Within indigenous cultures there has traditionally been an idea of looking after the environment. This is expressed in Australia in the phrase 'caring for country' – where 'country' includes land and sea territory. Allocating territory to coastal indigenous populations is one way to protect the oceans. It works on a local level but the benefits in preserving biological and cultural diversity spill over to global goods, preserving species and cultures to make the world a richer place for future generations.

OWNERSHIP OF INTERNATIONAL WATERS

If indigenous groups are granted some sea territory this is likely to be within the 12 nautical mile territorial zone of nation states, so it could come about through a change in national legislation as in the Blue Mud Bay case. It is worth remembering that the territorial zone determination and the setting of the EEZs are relatively recent phenomena coming into force only in 1994. The considerations in this book lead to the idea that this ownership regime should be unsettled in ways beyond according indigenous ownership to certain coastal sea territories.

The provision in the Law of the Sea which allows claims up to 350 nautical miles from the coast for the extraction of resources from the seabed didn't really get a grip until this century. New technologies that have allowed exploration under the sea ice and the partial melting of that ice in the Arctic for instance open up the possibility of extracting oil and gas under the Arctic Ocean. The Law did not take into account the difficulty in establishing a geological link between a coastal state and the seabed 350 nautical miles out. This is generating international conflict that is set to increase. There will be no easy resolution while the Law holds. In addition this provision in the Law of the Sea threatens the Antarctic Treaty. Now that states with territorial claims on the Antarctic landmass see that after 2048 they might be able to push a sea claim out to 350 nautical miles and harvest the resources on the seabed, the temptation to give up on the Antarctic Treaty will be great.

Where is the justice in being able to make a grab for the seabed resources over vast areas simply because of ownership of the land hundreds of miles away? Land-locked states receive no benefit from this part of the world's bounty. Where is the justice in extracting more and more resources that run a high risk of local environmental pollution in relatively pristine environments as a legacy for future generations? Where is the wisdom in pursuing the development of greater and greater oil and gas use so that more CO_2 is produced which will affect the atmosphere and kill off the oceans? Clearly there is no justice and there is no wisdom in going in this direction.

The arbitrariness of allocating 350 nautical miles of seabed to a coastal state built into the Law of the Sea should be exposed and questioned. However, we also need to look at the area from 12 to

200 nautical miles from the coast, namely most of the EEZ. Similar arguments about fairness and wisdom apply. This is not to say that resource extraction from the seabed should never occur. But mining resources that feed polluting industries must be curtailed and consideration should be given as to whether coastal states may retain sovereign rights over 200 nautical miles from their coast. An unfair dividing up of the world's resources needs more justification than simply the fact that it is built into the Law. This is especially the case when the extraction of these resources could well have a global impact. I am not suggesting that the sea territory beyond 12 nautical miles become high seas with the same free-for-all, unowned connotations as the present high seas. Rather the idea is to re-conceptualise this realm as belonging to everyone where everyone has an interest in its protection and a concern about the practices that threaten it. Major changes in ocean governance would be required, and I will return to this issue in the next section. The oceanic realm beyond 12 nautical miles could retain the name 'international waters', with greater justification than at present, where national claims to part of that territory are allowed.

There are further considerations that point to the view that the current legal regime governing EEZs and the high seas should be questioned. One is piracy. As I argued in Chapter 4, this is a growing problem impacting severely on international shipping. The attempts to address piracy have been hindered by problems about who should take responsibility for the pursuit and prosecution of pirates. International fleets are involved in trying to counter the problem in Somalia but they do not always act in an international interest, again as I pointed out in Chapter 4. They are operating in unowned waters so they can make their own decisions about who they will help and what they will do with the pirates if captured. As I noted there is a great deal of inconsistency, which leads to injustice and slows down the suppression of piracy.

By defining piracy as a high seas activity the Law of the Sea fails to connect with the bulk of attacks which mainly occur on ships in EEZs. If robberies or hijackings at sea take place within territorial waters then the coastal state should normally have jurisdiction. This was overridden in 2008 in Somalia by the UN, with the agreement of the Somali Government. Such a move is justifiable if the coastal state is incapable of dealing with the problem. The rest of the EEZ is presently open for all to navigate so there is no legal problem in the pursuit of sea robbers or hijackers in this zone. However the Law doesn't strictly cover their actions as they don't count as piracy. Why not then just say that sea robbers and hijackers are pirates if they conduct raids anywhere in the EEZ beyond the territorial sea as well as in the high seas? I think that this is the way to go. However the Law grants universal jurisdiction to piracy. As I argued in Chapter 4 this gives rise to conflict with the partial jurisdiction that coastal states have over the EEZ. If the EEZ out from 12 nautical miles is given over to some international body with full jurisdiction then this would resolve the conflicting legal standing of the area. It would also help to foster greater international co-operation in addressing piracy, not just in Somalia. If these waters are viewed as 'international waters' in the sense of being owned by the international community rather than belonging in some quasi-legal sense to the coastal state, then concerted anti-piracy moves might be initiated before the problems escalate to the level that they have in Somalia.

Thus to get a fairer distribution of undersea resources and to tackle piracy it could be desirable to redraw ocean borders. Abolishing the EEZ would mean that coastal states give up on the economic advantage accruing from their sovereign rights over this territory. Pursuing this direction, the biggest challenge concerns what to do about fisheries. In Chapter 5 I pointed out that the setting of the EEZ at 200 nautical miles followed on moves by Iceland and the European Economic Community to take control of that extent of territory out from their coasts in order to protect fish populations for their own exploitation. Such moves generated military conflict with other countries wanting to fish in these areas that eventually settled when the Law of the Sea took this change on board. More recently Canada has initiated moves to control fishing beyond the 200 mile EEZ which led to military posturing with Spain and the Straddling Stocks Agreement. This Agreement allows coastal states to enforce regional fishing agreements when they are violated by fishers in the high seas near the coastal state's EEZ border.

These changes in the regulatory regimes, while backed by law, still fall short of securing a just allocation of the world's fishing resources between states. Land-locked states miss out altogether. States which border highly productive waters may secure enormous benefits denied to others and those who are unable to stop foreign fishers entering their waters and taking fish illegally – as in Somalia, West Africa and Southeast Asia – fail to benefit from their EEZ. Coastal states are supposed to be looking after the conservation of the waters and marine life in the EEZ but where there is an unwillingness or inability to do so the fish populations will suffer. In Southeast Asia destructive fishing practices are severely depleting fish populations and habitats. In Somali waters hazardous and radioactive wastes have been dumped.[9] Fish living in these waters may be contaminated yet still marketed. Some countries have encouraged foreign fishers in the EEZs and received financial compensation but such fishing is likely to be unregulated and threaten the fish populations.[10]

The Straddling Stocks Agreement does something to protect fish on the high seas but only when they are swimming in areas close to states that have the capacity and desire to control illegal fishing, such as Canada. It is unenforceable in most areas or inapplicable, as in the Southern Ocean. I spent some time detailing the problems in curtailing illegal fishing of the migratory Patagonian Toothfish in that ocean despite having CCAMLR, a management body that was set up on excellent principles.

There are problems with migratory fish worldwide. Such fish may be of interest to states in different parts of the world; for example, tuna travel across the Atlantic from North American waters to the Mediterranean. Concern has been rising for some time over the depleted populations, with American fishers abiding by small quotas while larger quotas have been allowed in the Mediterranean along with illegal fishing taking place there.[11] The International Commission for the Conservation of Atlantic Tunas (ICCAT) has labelled the Mediterranean bluefin tuna fishery 'an international disgrace' even though the Commission is supposed to be guiding management of the species. ICCAT scientists claim that the population is on 'the brink of collapse'.[12] This prompted Spain and Japan to call for a ban on tuna fishing in the Mediterranean between May and June when the fish are spawning and particularly vulnerable, supporting a call already made by the US in 2007. Later in 2008, at a meeting of the IUCN, Italy and Japan voted in favour of a moratorium. ICCAT then decided to reduce the quotas, amongst other measures. However the drop in quota was only modest. Short-term business interests won out over any serious attempt to protect the tuna. According to a Greenpeace spokesman: 'The quota that ICCAT approved this week will basically take all of them ... The game's over.'[13]

Tuna in the Western and Central Pacific Ocean are also in trouble, with the possibility of a total collapse by 2030.[14] Here Japan has been accused of fishing over quotas. European fishing firms, Albacora, Calvopesca and Conservas Garavilla, sailing under flags of convenience from Venezuela, Panama, Ecuador and the Netherlands Antilles, are targeting tuna using illegal practices.[15]

Not only are some states disadvantaged by the present legal arrangements which affect fishing, we are also letting down future generations when our current practices oversee the demise of fish populations. There is clearly a need for greater international co-operation in addressing migratory fish populations or any high seas fish. However, the above reflections suggest a need for international control also in areas now designated as EEZs. Of course, simply dropping the border between EEZs and the high seas would lead to the battles of old over fishing grounds. A great deal more would need to be changed as suggested below.

Could whales and dolphins benefit from a change in sea borders? The IWC has a moratorium on whaling in all regions, but with some exceptions for aboriginal subsistence and scientific research. Moves to promote more non-lethal scientific studies of whales and continual condemnation by many other nations may eventually succeed in cutting off the scientific loophole. In the meantime, if most of the whale habitats are regarded as owned by all humanity, then there is the possibility of greater efforts being made to stop their slaughter and suffering. The IWC does not at present cover dolphins despite moves in that direction. There is a dire need to address the issue of the dolphin slaughter in Japan as well as the continual bycatch of dolphins in fishing operations globally. The slaughter occurs in the territorial sea around the Kii Peninsula. However a moratorium on the killing of dolphins imposed by the IWC would cover territorial seas as well as the rest of the ocean. It would also be an appropriate role for the IWC to work with fishing regulators to aim to eliminate dolphin bycatch. The IWC has expressed concern about the use of sonar and seismic technologies and their impact on whales. It is difficult for this message to amount to more than simply 'exercise caution' when states have jurisdiction over the areas where this testing is taking place. With a large reduction in this area, as suggested here, and a much more co-ordinated international management program, as will be outlined below, it would be easier for the IWC to have more influence on sonar and seismic testing by working with ocean regulators. If there was another international body with concern for the oceans it could also assist the IWC and individual states in their attempt to enforce the moratorium on commercial whaling that has so far proved difficult to enforce globally.

Enlightened policies to protect underwater cultural heritage for all humanity are encapsulated in the Convention on Underwater Heritage that came into force in January 2009, albeit with disappointingly few ratifications. As I pointed out in Chapter 3, inconsistencies with the Law of the Sea need to be addressed. On the one hand the Law supports the preservation of cultural heritage, on the other it allows for salvage of historic wrecks. There is also inconsistency between the Law and the Convention especially on the role of salvage, which the Convention sees as only applying to recent finds. Despite the few signatories to the Convention, the notion that underwater cultural heritage should be protected for the common benefit and not exploited for commercial ends is gaining momentum. One of the reasons put forward for not supporting the Convention by the US and UK is the requirement that states must preserve underwater cultural heritage in their waters for the benefit of humanity. This could be regarded as an onerous task. I believe that it is reasonable to require the input of the international community into the cost of such preservation measures if it is accepted that this action is for the benefit of all. These activities would be much easier to administer if the international teams were working outside of the areas where coastal states have claims and under the guidance of an international body. So in this field too there is reason to broaden the extent of the seas under international ownership.

Supposing the seas beyond 12 nautical miles do come under international ownership, what next? There would be a need for new institutions to ensure that this realm is one of international co-operation and benefit.

INTERNATIONAL OCEAN GOVERNANCE

A new international body allied to the UN should be established to take particular responsibility for the international waters proposed here. This brief should cover issues to do with the sea, seabed and oceanic life forms. Such an institution could be called the Oceans Council. All states whether coastal or not should be encouraged to join this Council as all states, collectively, own the area. Pro-rata payments by members should form the funding base. As Kaul argues: 'Public goods should be chosen and paid for by those who benefit from them – those with a global range of benefits should be paid for globally.'[16] The Council should be representative of all states and in order to achieve this outcome developing countries may need to be given assistance to enable full participation. It should be democratic with structures in place to prevent the domination of single countries or blocs of countries over others.

Another source of funding could be from ship registrations. The issue of flags of convenience needs to be addressed as the current system abets piracy, overfishing and pollution as well as unfair work practices. However, given that most ships are owned and operated by multinational companies it would not be possible to revert to the old system for all shipping where the nation of the ship's owner is the flag state. There is clearly a need for an international registry for shipping that makes the renewal of registration dependent on good practice. The Oceans Council would be ideally placed to take on the task of registering shipping. The fees could be used in part to place observers on boats to help curb the above practices.

The Oceans Council should make decisions on broad policy issues guided by a newly set up Intergovernmental Panel on the Oceans. The IUCN, represented by 84 states, has called for the establishment of such a panel to better inform policy making.[17] This could be based on and linked to the UN Intergovernmental Panel on Climate Change (IPCC). The IPCC is an information base for scientific findings about climate change. It does not conduct research but rather provides assessment of 'the latest scientific, technical and socio-economic literature produced worldwide relevant to the understanding of the risk of human-induced climate change, its observed and projected impacts and options for adaptation and mitigation'.[18] The IPCC does not make policy but provides a base for policy decisions by governments or climate change summits and negotiations under the Kyoto Protocol. It is part of the United Nations Environment Program (UNEP). Climate and the oceans are interwoven. The linkage between the two Panels would work both ways, since specific ocean effects such as acidification are an important indicator of atmospheric CO_2 levels, and the general statistics on rising temperatures are important information for ocean policy, for instance regarding sea-level rise.

The Law of the Sea would need to be changed to reflect the new ocean borders proposed here and to provide legal backing for the Oceans Council. The close links with shipping would allow the Council to take measures against ship-based pollution. However, there would also be a need to link with land-based organisations and scientists who are working on pollution of the ocean from the land in particular concerning agricultural run-off, the use of plastics and the dumping of hazardous waste in the sea. As some of these pollutants can have a global reach it is important for at least some of the efforts to counter pollution to come from an international organisation.

The Oceans Council could have a key role in countering piracy by working out strategies in advance of major problems arising, co-ordinating international efforts to address piracy, and setting up an international court to deal with prosecutions. According to the anti-piracy convention now in force the contracting governments are obliged to extradite or prosecute alleged offenders. I noted in Chapter 4 the inconsistent policies on prosecution that could be solved by having cases heard by a single court, which may however need to convene in various locations. Another issue that surfaced in 2008 is the reluctance of some countries to extradite captured pirates to countries that have the death penalty. The German Government suggested a special UN pirate court that would not accept such penalties. This may act as an incentive for more countries to be involved in the pursuit of pirates. However, if this court was part of the Oceans Council rather than the UN it would have the chance of being embedded in a more comprehensive framework for dealing with piracy.

The Oceans Council could form the broad guidelines for ocean resource management including taking account of climate change issues. The International Seabed Authority (ISA) has already proved a good initiative, and this independent organisation could be given powers and funding to expand its operations. The Authority has conducted workshops on environmental protection, drawing on experts in the field, and states that at present 'the understanding of deep-sea ecology is insufficient to allow conclusive risk assessment of large-scale mining'.[19] In addition, the environmental guidelines produced are only voluntary. There is a need to make these guidelines mandatory and to use the precautionary principle in all undersea mining. Decisions need to take into account the risk of mining fossil fuels not only for the immediate environment but also on a global level once exploited, bearing in mind guidelines that would be developed by the Oceans Council. The ISA is mentioned in the Convention on underwater cultural heritage as the body to receive notification of finds. It would be the appropriate body to deal with underwater cultural heritage in the expanded international waters. For this too the Authority would need extra funding and its powers would have to be increased if it is to play a role in co-ordinating international teams to work on shipwrecks and other items of cultural heritage.

An International Fishing Authority (IFA) could be established with the same overall principles as the ISA, namely, to work for the common benefit of humankind. If this aim applies to seabed resources and heritage in international waters there is no reason for it not to be applied to living resources also. The IFA should have the same type of management structure as the ISA. All states who are signatories to the Law of the Sea are members of the Authority and all are represented in the Assembly that sets the general policy. In the proposal being developed here these policy decisions would bear in mind advice from the Oceans Council. At present the executive arm of the ISA consists of 36 members who are elected by the Assembly for a rotating four-year term according to a formula intended to ensure the representation of all geographical blocs.[20] The IFA should also have close links with the Oceans Council but it should be responsible for developing specific policies on migratory fish populations concerning, for instance, moratoriums, sanctuaries, quotas, size limits and cruelty-free practices. It should also be responsible for issuing fishing licences to catch migratory fish populations if appropriate.

Regional bodies would deal with local fisheries acting on advice from the IFA. They should have a role in phasing out subsidies that now only serve to foster overfishing and in introducing subsidies for practices that promote the health of the oceans and oceanic life. There may be a need for compensation to be paid to developing countries where moratoriums or sanctuaries stop the possibility of fishing. In 2006, Kiribati created the world's third largest marine reserve. This will protect more than 120 species of coral and 520 species of fish inside 184,700 square kilometres. Compensation and an endowment to pay for the park's management are covered by the New England Aquarium in Boston and Conservation International, a non-government organisation.[21] In a welcome parting gift, US President George W. Bush added another 505,000 square kilometres to marine national parks around US-controlled islands in the Pacific Ocean. Most commercial fishing will be banned and limits placed on other fishing.[22] These are important initiatives but there are likely to be more along these lines if there is some sort of regular institutional support that could, for instance, be provided by regional fishing bodies or the IFA.

The regional authorities could take over the issuing of fishing licences with fees on an escalating scale depending on the size of the operation. The licence programme should pay particular attention to the needs of coastal communities whose livelihood has been threatened by large fishing operations. The regional authorities should represent not only the coastal states but also land-locked states in the region given that the international waters are just as much 'theirs' as anyone else's.

Some of the funding could be used to place observers on the larger fishing trawlers to ensure compliance with the regional fishing authority's policies. The regional fishing authorities should have the power to deny licences to unsustainable fisheries or to revoke them. Fishers' practices such as those followed in Iceland could be brought in, whereby access and profits are shared and there is a common interest in not depleting the fish populations or wrecking the ecology. These regional bodies could take over the work previously done by organisations such as CCAMLR, NAFO and ICCAT that began well but fell too much under the influence of 'resource extraction' thinking. The new fishing authorities would need to have a dual economic/conservation orientation. Fish are crucial to the protein requirements of over 1 billion people, but without conservation policies and actions these needs won't be met in the future. Careful strategies are required to preserve the oceans and oceanic life in order to have a marine economy at all.

Information about sustainable fisheries should be made available to consumers through organisations such as the Marine Stewardship Council, an independent non-profit body based in London, now operating in 36 countries and rapidly expanding. This Council was set up in response to the collapse of the cod populations in Canada. Standards for sustainable fishing have been developed and a labelling programme ensures that the fish or seafood can be traced back to a sustainable fishery.[23]

The IFA observers could monitor shipping practices, such as the discarding of nets and gear, that cause problems to marine life. Engaging in these practices could lead to revocation of a fishing licence. Some of the funding from registration fees could be used to tackle pollution from ships and provide incentives for good practice. International co-operative efforts to clean up surface garbage should be encouraged, particularly by the regional fishing authorities, along the lines of coastal 'clean-up' campaigns such as the very successful 'Clean-up Australia' and the new 'Clean-up the World' campaign now working in partnership with UNEP.

A representative from a non-government body in both the ISA, IFA and the regional authorities could reveal imbalances in participation and representation and assist in the transparency of the deliberations. Support for developing nations to attend meetings might be necessary. The ISA and the IFA would need a funding base beyond the issuing of licences to mine or to fish. As they both need to take conservation seriously to avoid irreparable damage to the oceans and the life therein, and as these aims concern all people, a pro-rata funding arrangement should be worked out.

IMPLEMENTATION OF A NEW OCEAN MANAGEMENT REGIME

The biggest change suggested here is the re-drawing of ocean boundaries. If that occurs then new strategies for dealing with piracy, fishing, undersea resources and heritage will have to follow, as current policies rely heavily on legal claims over sea areas by coastal states.

How the oceans are divided up is determined by the Law of the Sea. This Law can be changed by a UN Convention receiving a certain number of ratifications. Changes from the Convention that came into force in 1994 are unlikely to occur without a shift in material conditions. I suggest that we are moving into an era where such changes are occurring. They include:

1. The escalation of military threats between states making ownership claims on the Arctic Ocean, and political moves to make such claims on the Southern Ocean in the near future.
2. The growing awareness of the fragility of the Arctic and what might be lost with oil and gas exploitation.
3. Increasing conflict over the ownership of shipwrecks found in the high seas and EEZs.
4. The rise in piracy on a large scale in the waters of states unwilling or unable to exercise control together with slow reactions by the international community.
5. The depletion of the world's fish populations to dangerous levels with no clear way of turning this situation around under the current ownership regime.
6. The mounting threat to cetaceans, even endangered populations, with coastal states unwilling to curtail dangerous testing and undersea exploitation in their EEZs.
7. A growing recognition of the value to humankind of the preservation of cultural diversity and a need to grant indigenous groups sea territory.
8. A concern for the future of the planet and the impact of rising CO_2 levels not only on the land but also on the oceans.

International links are already beginning to tackle the issues raised in this book: Australia and France have joined forces to combat illegal fishing in the Southern Ocean; Malaysia, Indonesia and Singapore co-ordinate anti-piracy moves in the Straits of Malacca; a very large international contingent is employed in fighting Somali piracy; and the US, Spain, Italy and Japan are working co-operatively to prevent the demise of tuna. The IUCN, UNEP, the Marine Stewardship Council, many marine mammal organisations and others play a vital role in arousing global concern for the oceans.

I am hopeful that these material conditions, together with the development of international co-operation to protect the oceans, will be sufficient to lead to a new UN Convention on the Law of the Sea in which the ownership regime is changed. The prospect of a vast part of the world coming under communal ownership and responsibility has an intrinsic appeal. It is a move that has a good chance of setting in train a range of changes that might bring about greater equity between states and between generations.

Notes

(Website addresses last accessed: June 10, 2009)

INTRODUCTION

1. BBC News, August 2, 2007, http://news.bbc.co.uk/2/hi/europe/6927395.stm
2. CNN.com/world, August 2, 2007, http://edition.cnn.com/2007/WORLD/europe/08/02/arctic.sub.reut/index.html
3. Ibid.
4. R. Carson, *The Edge of the Sea*, Boston: Houghton Mifflin, 1955, p. 250.

1 FREEDOM OF THE SEAS

1. Alexander McCall Smith, *The Sunday Philosophy Club*, London: Little Brown, 2004.
2. T. W. Fulton, *The Sovereignty of the Sea*, Edinburgh: Blackwell and Sons, 1911, pp. 340–1.
3. Ibid., pp. 3–4.
4. Ibid. p. 4.
5. F. Braudel, *The Mediterranean and the Mediterranean World in the Age of Philip I*, Vol. 2, trans. Stan Reynolds, London: Collins, 1973, p. 783.
6. P. Gosse, *The History of Piracy*, New York: Tudor Pub. Co., 1932, p. 11.
7. Ibid.
8. Ibid., pp. 14–15.
9. Ibid., p. 18.
10. Ibid., p. 27.
11. U. Klausman, M. Meinzerin and G. Kuhn, *Women Pirates and the Politics of the Jolly Roger*, trans. Tyler Austin and Nicholas Levis, Montreal: Black Rose Books, 1997, pp. 97–8.
12. Gosse, *The History of Piracy*, pp. 19–20.
13. Ibid., p. 24.
14. Ibid., pp. 28–9.
15. Braudel, *The Mediterranean*, p. 882.
16. Ibid.
17. Ibid.
18. Ibid., p. 875.
19. Ibid.
20. Ibid., p. 878.
21. Ibid., pp. 886–7.
22. Ibid., p. 867.
23. Gosse, *The History of Piracy*, p. 54.
24. Braudel, *The Mediterranean*, p. 886.

25. Adriaan de Jong (translator), *Spice Adventurers: The Original Logs of the Ship Gelderland, 1601–1603*, Canberra: Replica Foundation Inc., 2002 (Duyfken, 1606), p. 4.
26. Ibid.
27. M. S. Ball, *Lying Down Together: Law, Metaphor and Theology*, Wisconsin: University of Wisconsin Press, 1985, p. 39.
28. John Villiers, 'The Estado da India in Southeast Asia', in Malyn Newitt (ed.), *The First Portuguese Colonial Empire*, Exeter: University of Exeter, 1986, p. 42.
29. de Jong, *Spice Adventurers*, p. 5.
30. Ibid.
31. Ibid.
32. Ball, *Lying down Together*, p. 146.
33. Gosse, *The History of Piracy*, p. 58.
34. Braudel, *The Mediterranean*, p. 866.
35. Ibid., pp. 865–6.
36. Ibid., p. 890.
37. Ibid., p. 868.
38. Ibid.
39. Ibid., p. 878.
40. Ibid., p. 884.
41. Ibid., p. 885.
42. Cornelius von Bykershoek, *A Treatise on the Law of War*, trans. Cornelius von Bykershoek, Philadelphia: American Law, 1810 (1744), p. 127.
43. Ibid., pp. 131–2.
44. J. E. Thomson, *Mercenaries, Pirates, and Sovereigns: State-building and Extra-territorial Violence in Early Modern Europe*, Princeton: Princeton University Press, 1994, p. 45.
45. C. Richardson, *Pirates*, Sydney: Scholastic, 2004, p. 6.
46. Quoted in M. Weatherly, *Women Pirates: Eight Stories of Adventure*, Greensboro: Morgan Reynolds Inc., 1998, p. 27.
47. Ibid., p. 32.
48. A. Gill, *The Devil's Mariner: A Life of William Dampier, 1651–1715*, London: Michael Joseph, 1997, p. 17.
49. A. S. George, *William Dampier in New Holland: Australia's Natural Historian*, Hawthorn, Victoria, Australia: Bloomings Books, 1999, p. 17.
50. Ibid., p. 18.
51. A. Gentili, *De Jure Belli Libri Tres*, Vol. 2, trans. John C. Rolfe, Oxford: The Clarendon Press, 1933, p. 124.
52. J. B. Scott (ed.), Hugo Grotius, *The Freedom of the Seas or the Right which Belongs to the Dutch to Take Part in the East Indian Trade*, trans. Ralph Van Deman Magoffin, New York: Oxford University Press, 1916 (1608).
53. Ibid., pp. 7, 98.
54. Ibid., p. 8.
55. Ibid.
56. Ibid., p. 10.
57. Ibid., pp. 22–45.
58. Ibid., pp. 40–1.
59. Ibid., p. 41.
60. Ibid., p. 39.

61. Ibid., pp. 24–8, 32, 38.
62. Ibid., p. 38.
63. Ibid., p. 43.
64. Ibid., p. 45.
65. Ibid.
66. Ibid., p. 46.
67. Ibid., p. 47, 50.
68. Ibid., p. 52.
69. Ibid., p. 55.
70. Ibid., pp. 60, 75.
71. W. Welwod, 'On the Community and Property of the Seas', in D. Armitage (ed.), *The Free Sea*, trans. Richard Hakluyt, Indianapolis: Liberty Fund, 2004 (1872), p. 74.
72. Ibid.
73. H. Grotius, 'Defence of Chapter V of the *Mare Liberum*', in Armitage (ed.), *The Free Sea*, p. 83.
74. J. Selden, *Mare Clausum: Of the Dominion, Or, Ownership of the Sea*, New Jersey: The Lawbook Exchange Ltd, 2004 (1652), p. 175.
75. Ibid., pp. 135–8.
76. Ibid., p. 171.
77. Ibid., p. 27.
78. Ibid., p. 108.
79. Ibid., unnumbered.
80. Ibid.
81. Fulton, *The Sovereignty of the Sea*, pp. 340–1.
82. C. R. Boxer, *The Dutch Seaborne Empire 1600–1800*, London: Hutchinson, 1965, p. 97.
83. Fulton, *The Sovereignty of the Sea*, p. 341.
84. Ibid.
85. Scott, *The Freedom of the Seas*, p. ix.
86. Ibid., p. 29.
87. Ibid., p. 30.
88. Ibid.
89. Ibid., p. 32.
90. Ibid., pp. 35, 37.
91. Ibid., p. 37.
92. Ibid.
93. Ball, *Lying Down Together*, p. 41.
94. *The United Nations Convention on the Law of the Sea with Index and Final Act of the Third United Nations Conference on the Law of the Sea* (LOSC), New York: United Nations, 1983.
95. Scott, *The Freedom of the Seas*, p. 32.
96. LOSC, Article 19.
97. D. R. Rothwell, 'UNCLOS as an Instrument of Maritime Confidence Building Measures in East Asia', in D. K. Kim, C. Park and A. Lee (eds), *UN Convention on the Law of the Sea and East Asia*, Seoul: Seoul Press, 1996.
98. S. Bateman and A. Bergin, *Sea Change: Advancing Australia's Ocean Interests*, Canberra: Australian Strategic Policy Institute, 2009, p. 43.
99. *The Australian*, December 15, 2004.
100. Bateman and Bergin, *Sea Change*, p. 11.

101. 'Indonesia Refuses Australian Maritime Information Zone', December 24, 2004, http://www.tempointeractive.com
102. ABC News, December 16, 2004, http://www.abc.net.au/news/stories/2004/12/16/1266479.htm
103. *Spiegel Online International,* December 18, 2008, www.spiegel.de/international/europe/0,1518,597340,00.html
104. S. Earle, *Sea Change: A Message of the Oceans,* New York: G. P. Putnam's Sons, 1995, p. xii.
105. James J. Titus, 'Policy Implications of Sea-level Rise: The Case of the Maldives', *Proceedings of the Small States Conference on Sea-level Rise,* Malé, Republic of Maldives, November 14–18, 1989, p. 4.
106. United Nations Development Program: Human Development Report, 2007/2008, 'Fighting Climate Change: Human Solidarity in a Divided World', Chapter 2, p. 93, http://hdr.undp.org/en/reports/global/hdr2007-2008
107. Ibid., p. 91.
108. *Spiegel Online International,* November 18, 2008, http://www.spiegel.de/international/world/0,1518,591164,00.html
109. Planet Ark News, November 10, 2006, http://www.planetark.org/dailynews-story.cfm?newsid=38918&newsdate=10Nov2006
110. 'Seychelles Identify with Fate of Pacific Islands', December 9, 2008, http://www.islandsbusiness.com/news/index_dynamic/containerNameToReplace=MiddleMid
111. 'Climate Change is a Battle for Existence in the Maldives', November 25, 2008, http://www.google.com/hostednews/afp/article/ALeqM5iJltPXP3dS8I-UoPulsUs123tStWg.

2   UNDERWATER NON-LIVING RESOURCES

1. M. Sommerkom and N. Hamilton (eds), *Arctic Climate Impact Science: An Update Since ACIA,* Oslo: WWF International Arctic Programme, 2008.
2. *The United Nations Convention on the Law of the Sea* (LOSC).
3. Annica Carlsson, 'The US and UNCLOS III – The Death of the Common Heritage of Humankind Concept?', *Maritime Studies,* No. 95 (July–August 1997), p. 2.
4. International Seabed Authority: Convention and Agreement, www.isa.org.jm.en/home
5. David B. Sandalow, 'Law of the Sea Convention: Should the US Join?', 2004, http:www.brookings.edu/papers/2004/08eneregy_sandalow.aspx?p=1
6. *The Washington Note,* May 14, 2008, http://www.thewashingtonnote.com/archives/002128.php
7. Sandalow, 'Law of the Sea Convention', p. 6.
8. LOSC, Article 76.
9. LOSC, Article 77.
10. LOSC, Article 76.
11. Commonwealth of Australia, *The Antarctic Treaty System,* Canberra: Australian Government Publishing Service, 1986.
12. *Guardian,* October 17, 2007.
13. *Spiegel Online International,* July 5, 2008, http://www.spiegel.de/international/world/0.1518,551807.00.html
14. Commonwealth of Australia, *The Antarctic Treaty System,* p. 20.

15. Gary Brown, 'The Parliament of the Commonwealth of Australia: The Future of the Antarctica', *Current Issues Paper 4*, 1987–8, p. 1.
16. *Guardian*, October 17, 2007.
17. Environmental News Network, August 23, 2007, http://www.enn.com/top_ stories/article/22219
18. *Sydney Morning Herald*, August 13, 2007.
19. Environment News Service, February 11, 2008, http://www.ens_newwire. com/ens/feb2008/2008-02-00-01.asp
20. J. L. Hansen, 'Ice, Shrimps and Oil Worth Billions', *Ministry of Foreign Affairs of Denmark*, February, 2008, http://www.netpublikationer.dk/um/8976/html/ chapter07.htm
21. *Sydney Morning Herald*, June 13, 2008.
22. CBC News, 'Harper Announces Northern Deep-sea Port, Training Site', August 11, 2007.
23. Speech by Commissioner Joe Borg at the Conference, 'Common Concern for the Arctic', Ilulissaat, Greenland, September 9, 2008.
24. *The European Weekly*, December 12, 2008, http://www.neurope.eu/articles/ 90700.php
25. MSN, 'Canada Requires Ship Registration in Arctic', August 27, 2008, http:// www.msnbc.msn.com/id/26419116
26. BBC News, January 30, 2008, http://new.bbc.co.uk/2/hi/in_depth/629/629/ 7214857.stm
27. Norwegian Chairmanship, *Arctic Council: Oil and Gas Assessment*, 2008, http://www.amap.no/oga
28. Ibid., Chapter 4, pp. 10–12.
29. Ibid., Chapter 5, p. 98.
30. Ibid., Chapter 5, p. 97.
31. Ibid., Chapter 5, p. 98.
32. Ibid., Chapter 4, p. 16.
33. Ibid., Chapter 5, p. 25.
34. Charles H. Peterson, Stanley D. Rice, Jeffrey W. Short, Daniel Esler, James L. Bodkin, Brenda E. Ballachey, David B. Irons, 'Long-Term Ecosystem Response to the *Exxon Valdez* Oil Spill', *Science*, Vol. 302, No. 5653 (December 2003), p. 2082.
35. Norwegian Chairmanship, *Arctic Council*, Chapter 4, p. 29.
36. Ibid., Chapter 4, p. 19.
37. Ibid., Chapter 4, p. 29.
38. Ibid., Chapter 4, p. 20.
39. Ibid., Chapter 4, p. 14.
40. Ibid., Chapter 5, pp. 5–6.
41. Ibid., Chapter 5, p. 16.
42. Ibid., Chapter 5, pp. 163–4.
43. Ibid., Chapter 4, pp. 22–8.
44. Ibid., Chapter 4, p. 64.
45. *Speigel Online International*, January 23, 2008, http://www.spiegel.de/ international/world/0,1518,530454,00.html
46. Sommerkom and Hamilton, *Arctic Climate Impact Science*; Carolyn Symon, L. Arris and Bill Heal (eds), *Arctic Climate Impact Assessment* (ACIA), Cambridge: Cambridge University Press, 2005.
47. Sommerkom and Hamilton, *Arctic Climate Impact Science*, p. 1.

48. Symon et al., *Arctic Climate Impact Assessment.*
49. Sommerkom and Hamilton, *Arctic Climate Impact Science*, p. 7.
50. Ibid., p. 51.
51. Ibid., p. 67.
52. Ibid., p. 53.
53. Ibid., p. 68.
54. Ibid., p. 68.
55. B. M. Jenssen, 'Endocrine-disrupting Chemicals and Climate Change: A Worst-case Combination for Arctic Marine Mammals and Seabirds?', *Environmental Health Perspectives*, Vol. 114, Supplement 1 (2006), pp. 76–80.
56. Sommerkom and Hamilton, *Arctic Climate Impact Science*, p. 59.
57. Ibid., p. 58.
58. I. Stirling, 'Reproductive Rates of Ringed Seals and Survival of Pups in Northwestern Hudson Bay, Canada, 1991–2000', *Polar Biology*, Vol. 28 (2005), pp. 381–7.
59. Sommerkom and Hamilton, *Arctic Climate Impact Science*, p. 58.
60. Ibid., p. 10.
61. Planet Ark News, May 16, 2008, http://www.planetark.com/avantgo/dailynewsstory.cfm?newsid=48369
62. Sommerkom and Hamilton, *Arctic Climate Impact Science*, p. 58.
63. Ibid., p. 89.
64. C. Nellemann, S. Hain and J. Alder (eds), *In Dead Water: Merging of Climate Change with Pollution, Over-harvest, and Infestations in the World's Fishing Grounds*, GRID-Arendal, Norway: United Nations Environment Programme, 2008, p. 35.
65. ScienceAlert, May 6, 2008, http://www.sciencealert.com.au/features/2008060S-17277-2.html
66. ScienceAlert, November 22, 2007, http://www.sciencealert.com.au/opinions/20072211-16629-2.html
67. Ibid.
68. C. Veron and W. Howard, 'Ocean Acidification – The BIG global Warming Story', *Catalyst*, September 13, 2007.
69. Science News, October 19, 2007, http://www.sciencedaily.com/releases/2007/10/071017102133.htm
70. *Sydney Morning Herald*, November 12, 2008.
71. UNDP Human Development Report, 2007/2008, Chapter 2, p. 107, http://hdr.undp.org/en/reports/global/hdr2007-2008
72. Fridtjof Nansen, *Farthest North: The Norwegian Polar Expedition, 1893–1896*, London: George Newnes Ltd, 1898, Vols 1 and 2.
73. Ibid., Vol. 1, p. 282.
74. Ibid., p. 284.
75. B. Lopez, *Arctic Dreams: Imagination and Desire in a Northern Landscape*, London: The Harvill Press, 1986.

3 UNDERWATER CULTURAL HERITAGE

1. W. Shakespeare, *The Tempest*, Cambridge: Cambridge University Press, 1969, p. 6.
2. J. Clottes, Courtin J. Beltrán and H. Cosquer, 'The Cosquer Cave on Cape Morgiou, Marseilles', *Antiquity*, Vol. 656 (1992), pp. 583–98.

3. J. Empereur, 'Diving on a Sunken City', *Archaeology*, Vol. 52, No. 2 (March–April 1999), pp. 36–46.

4. J. G. Royal, 'Discovery of Ancient Harbour Structures in Calabria, Italy, and Implications for the Interpretation of Nearby Sites', *International Journal of Nautical Archaeology*, Vol. 37, No. 1 (2008), pp. 49–66.

5. *ScienceDaily*, March 15, 2008, http://www.sciencedaily.com/releases/ 2008/03/ 080311120621.htm

6. Robert L. Hohlfeder and R. Vann, 'Cabotage at Aperlae in Ancient Lycia', *International Journal of Nautical Archaeology*, Vol. 29, No. 1 (2000), pp. 126–35.

7. Djutjadjutja Munungurr, 'Mana at Lutumba', in Buku-Larrngay Mulka Centre, *Saltwater: Yirrkala Bark Paintings of Sea Country – Recognising Indigenous Sea Rights*, Northern Territory, Australia: Buku-Larrngay Mulka Centre, 1999, p. 29.

8. Yangarriny Wununmurra, 'Galkama at Garraparra', in Buku-Larrngay Mulka Centre, *Saltwater*, p. 40.

9. Buku-Larrngay Mulka Centre, *Saltwater*, 1999, p. 104.

10. Richard A. Gould, *Archaeology and the Social History of Ships*, Cambridge: Cambridge University Press, 2000, pp. 98–105.

11. Ibid., p. 122.

12. Claire Peachey, 'Uluburun: Continuing Study of the Uluburun Shipwreck Artifacts', Institute of Nautical Archaeology, Bodrum, Turkey, undated, http:// diveturkey.com/inaturkey/uluburun/conservation.htm

13. The records of shipwrecks are now very extensive but the largest are not internet-based. There are two key sources of information. Harald Bolten holds a comprehensive shipwreck database and will provide details about particular wrecks if the name of the ship is sent to him: Global Shipwreck Database: http://members.aol.com/HBbolten/Database.htm. The Northern Shipwrecks Database: www.northernmaritimeresearch.com, has over 100,000 records of North American wrecks. The information is available by sending a letter with a small fee.

14. Anastasia Strati, 'Greece', in Sarah Dromgoole (ed.), *The Protection of Underwater Cultural Heritage: National Perspectives in Light of the UNESCO Convention 2001*, Leiden: Martinus Nijhoff Publishers, 2006, p. 125.

15. Sarah Dromgoole, 'United Kingdom', in Dromgoole, *The Protection of Underwater Cultural Heritage*, p. 316.

16. Ibid.

17. Sarah Dromgoole, 'United Kingdom', in Sarah Dromgoole (ed.), *Legal Protection of the Underwater Cultural Heritage: National and International Perspectives*, London: Kluwer Law International, 1999, p. 186.

18. Ibid. p. 186.

19. Dromgoole, 'United Kingdom' (2006), p. 320.

20. Jeremy Green, *Maritime Archaeology: A Technical Handbook*, second edition, Oxford: Elsevier, 2004, p. 403.

21. *The Daily Mail*, June 13, 2007, http://www.dailymail.co.uk/pages/live/articles/ news/worldnews.htm?in_article_id=461791

22. *The Independent*, April 10, 2008.

23. Gould, *Archaeology and the Social History of Ships*, p. 2.

24. N. C. Dobson, 'A Message From "the People's Archaeologist"', *Odyssey Marine Exploration*, August 4, 2006, http://www.shipwreck.net/ou/ou-aug04-06shtml

25. Gould, *Archaeology and the Social History of Ships*, p. 317.

26. *The United Nations Convention on the Law of the Sea* (LOSC).

27. Anastasia Strati, *The Protection of the Underwater Cultural Heritage: an Emerging Objective of the Contemporary Law of the Sea*, The Hague: Nijhoff Pubs, 1995, pp. 185–6.

28. Ibid., p. 269.

29. Ole Varmer and C. M. Blanco, 'United States of America', in Dromgoole (ed.), *Legal Protection of the Underwater Cultural Heritage*, pp. 205–21; Ole Varmer, 'United States of America', in Dromgoole (ed.), *The Protection of Underwater Cultural Heritage*, pp. 356–9.

30. Varmer and Blanco, 'United States of America', pp. 205–6.

31. R. J. Ella, 'US Protection of Underwater Cultural Heritage Beyond the Territorial Sea: Problems and Prospects', *International Journal of Nautical Archaeology*, Vol. 29 No. 1 (2000), p. 45.

32. Jay Cashmere, 'Atocha Wreck', October 10, 2007, http://www.wptv.com./content/segments/uww/story/aspx?content_id=04e484bo-cfd9-44d8-a529-8dc7lal4e192

33. Pat Clyne, 'Spain Expressly Abandon's [sic] its Shipwrecks!', undated, http://www.imacdigest.com/spainarticle.html

34. Archaeology Program, National Park Service, US Department of the Interior, http://www.nps.gov/archaeology/sites/npSites/assateague.htm

35. Jim Sinclair, *Deliver the Deliverance?*, 2007, http://www.imacdigest.com/deliverance.html

36. *Guardian Unlimited*, June 2003, http://www.buzzle.com/editorials/1-6-2003-33175.asp

37. Gould, *Archaeology and the Social History of Ships*, p. 322.

38. Green, *Maritime Archaeology*, p. 402.

39. Gould, *Archaeology and the Social History of Ships*, p. 2.

40. UNESCO Convention on the Protection of Underwater Cultural Heritage 2001, http://portal.unesco.org/en/ev.php-URL_ID=13520&URL_DO=DO+TOPIC&URL_SECTION=201l.html

41. Ibid., pp. 1–2.

42. Ibid., p. 2.

43. Ibid., p. 2.

44. I. Seidl-Hohenveldern, 'Flags of Convenience', in B. Vukas (ed.), *Essays on the New Law of the Sea*, Zagreb: Faculty of Law, University of Zagreb, 1990, p. 301.

45. Sinclair, *Deliver the Deliverance?*, p. 2.

46. ABC Online, May, 2007, p. 1, http://www.abc.net.au//news/newsitems/200705/s1927535.htm

47. 'Odyssey Provides "*Black Swan*" Shipwreck Information Update', *Odyssey Marine Exploration*, May 2007, http://www.shipwreck.net/pr135.html

48. Spain News and Information in English, May 8, 2008, http:www.typicallyspanish.com/news/publish/article_16411,sgtnk

49. ABC News, May 9, 2008, http://www.abc.net.au/news/stories/2008/05/09/2239750.htm

50. Strati, *The Protection of Underwater Cultural Heritage*, p. 243.

51. *New York Times*, May 12, 2008, p. 1.
52. LOSC, Article 149.
53. Anthony Firth, 'The UNESCO Convention for the Protection of the Underwater Cultural Heritage: Proceedings of the Burlington House Seminar, October 2005 – by the Joint Nautical Archaeology Policy Committee', *International Journal of Nautical Archaeology*, Vol. 37, No. 1 (2008), p. 3.
54. Ibid.
55. 'United Kingdom and Odyssey Sign Partnering Agreement for Sussex Shipwreck', October, 2002, p. 1, http://findarticles.com/p/articles/mi_mOEIN/is_2002_Oct_7/ai_02536241
56. CDNN-CYBER DIVER News Network, October 24, 2007, http://www.cdnn.info/news/industry/i071024.html
57. Robert C. Blumberg, 'International Protection of Underwater Cultural Heritage', *US Department of State*, March 2005, p. 3, http://www.state.gov/g/loes/rls/rm/51256.htm

4   MODERN PIRACY AND TERRORISM ON THE SEA

1. Quoted in W. Langewiesche, *The Outlaw Sea: A World of Freedom, Chaos, and Crime*, New York: North Point Press, 2004.
2. Robert I. Burns, *Muslims, Christians and Jews in the Crusader Kingdom of Valencia*, New York: North Point Press, 2004.
3. Langewiesche, *The Outlaw Sea*, pp. 47–9, 53.
4. Ibid., p. 52.
5. Ibid., pp. 52–9.
6. International Chamber of Commerce (ICC), International Maritime Bureau, 'Piracy and Armed Robbery Against Ships', *Annual Report*, January 1–December 31, 2004, p. 4.
7. ICC, *Report*, 1 January–September 30, 2009, pp. 6–7.
8. ICC, *Annual Report*, 2004, p. 4.
9. ICC, *Annual Report*, 2008, p. 5; ICC, *Report*, 2009, p. 6.
10. ICC, *Annual Report*, 2008, p. 5.
11. ICC, *Report*, 2009, p. 6.
12. Ibid.
13. LOSC, Article 101.
14. LOSC, Article 105.
15. Langewiesche, *The Outlaw Sea*, p. 71.
16. Ibid., p. 76.
17. Ibid.
18. International Maritime Organisation, *IMO Misc.4*/Circ 93, November 10, 2006.
19. ICC, *Annual Report*, 2008.
20. Ibid.
21. ICC, *Annual Report*, 2008, p. 13.
22. ICC, *Report*, 2009, p. 14.
23. ICC, *Annual Report*, 2008, p. 13.
24. ICC, *Report*, 2009, p. 14.
25. BBC News, April 28, 2009.
26. D. Dillon, 'Maritime Piracy: Defining the Problem', *SAIS Review*, Vol. 25, No. 1 (2005), p. 159.

27. Ibid., p. 160.
28. ICC, *Annual Report*, 2008, p. 23.
29. Ibid., p. 4.
30. Peter Chalk, 'Grey-Area Phenomena in Southeast Asia: Piracy, Drug Trafficking and Political Terrorism', Canberra: Strategic and Defence Studies Centre, ANU, 1997.
31. J. S. Burnett, *Dangerous Waters: Modern Piracy and Terror on the High Seas*, New York: Plume, 2003, p. 157.
32. Chalk, 'Grey-Area Phenomena' pp. 18–19.
33. World News Network, November 19, 2009, http:wn.com/view/2009/11/10/Somali_Pirates_seize_arms
34. Burnett, *Dangerous Waters*, p. 85.
35. Ibid., p. 157.
36. Ibid., p. 100.
37. *New York Times*, October 1, 2008.
38. ICC Commercial Crime Services, January 28, 2009, www.icc-ics.org
39. Information for Blue-water Sailors, http://www.yacht-piracy.org
40. C. Torchia, 'Somali Pirates Hone their Tactics', May 26, 2009, http://www.google.com/hostednews/ap/article/ALeqM5goVMzJStGaVB
41. Burnett, *Dangerous Waters*, pp. 218–19.
42. S. Robinson and Xan Rice, 'In Peril on the Sea', *Time*, Asian Edition, November 7, 2005, p. 1.
43. S. Stone, 'Pirates Attack 2 Navy Warships from Norfolk in Indian Ocean', *The Virginian-Pilot*, March 19, 2006, http://hamptonroads.com/story.cfm?story=101669&ran=240555
44. ICC, *Annual Report*, 2007, pp. 6, 81–6.
45. ICC Commercial Crime Services, *Weekly Piracy Report*, February 12–18, 2008, http://www.icc.ccs.org/prc/piracyreport.php
46. BBC News, November 18, 2009.
47. UN Security Council, *SC/9344*, Resolution 1816, 2008.
48. Russia Today News, November 21, 2008, http://www.russiatoday.com/news/news/33556
49. *Hindustan Times*, February 27, 2009, epaper.hindustantimes.com
50. 'Wrong Signals', May 7, 2009, Economist.com
51. ABC News, April 18, 2009, http://www.abc.net.au/news/stories/2009/04/18/2546583.htm
52. *The Economist*, May 9, 2009.
53. Fox News, November 26, 2008.
54. BBC News, April 28, 2009.
55. ICC, *Annual Report*, 2008, p. 9; ICC, *Report*, 2009, p. 5.
56. ICC, *Annual Report*, 2004, p. 12; ICC, *Annual Report*, 2006, p. 14; ICC, *Annual Report*, 2008, p.16, ICC, *Report*, 2009, p. 16.
57. Information for Blue-water Sailors, http:www.yacht-piracy.org
58. http://www.noonsite.com
59. Information for Blue-water Sailors, http:www.yacht-piracy.org
60. Ibid.
61. 'Pirates Attack Swedish yacht in Venezuela', March 31, 2001, http://www.noonsite.com
62. 'Yet Another Fatal Attack on Yacht in Venezuela', June 30, 2004, http://www.noonsite.com

63. ICC, *Annual Report*, 2008, p. 37.
64. Ibid., p. 31.
65. *Guardian Unlimited*, April 11, 2009, guardian.co.uk
66. *Sydney Morning Herald*, November 2, 2009.
67. A. Cassese, *Terrorism, Politics and the Law: The Achille Lauro Affair*, trans. Jennifer Greenleaves, Cambridge: Polity, 1989, p. 6.
68. Ibid.
69. Ibid., p. 26.
70. Ibid., p. 28.
71. Ibid., p. 32.
72. Ibid., p. 34.
73. Ibid., p. 38.
74. Ibid., p. 99.
75. International Maritime Organisation, 'Convention for the Suppression of Unlawful Acts Against the Safety of Maritime Navigation, 1988', http://www.imo.org/Conventions
76. Ibid., p. 2.
77. James Bamford, *Body of Secrets*, London: Arrow Books, 2002.
78. L. Wright, *Looming Tower*, New York: Knopf, 2006, pp. 322–31.
79. M. C. W. Chong, *Suppressing Piracy jure gentium*, Ph.D. dissertation, University of Sydney, 2007.

5   THE FISHING WARS

1. M. Kurlansky, *Cod: A Biography of the Fish That Changed the World*, New York: Penguin, 1998.
2. G. Thorlacius Johannesson, 'How "Cod Wars" Came: The Origins of the Anglo–Icelandic Fisheries Dispute, 1958–61', *Historical Research*, Vol. 77, No. 198 (November 2004), pp. 543–74.
3. D. Kassebaum, 'Cod Dispute Between Iceland and the United Kingdom', *ICE Case Studies*, Case Number 7, 1997, http://www.american.edu.TED/ice/CODWAR.HTM
4. G. Hellmann and B. Herborth, 'Fishing in the Mild West: Democratic Peace and Militarised Interstate Disputes in the Transatlantic Community', *Review of International Studies*, Vol. 34 (2008), pp. 481–506.
5. Quoted in Kurlansky, *Cod*, p. 167.
6. Kassebaum, 'Cod Dispute'.
7. C. Glover, *The End of the Line: How Overfishing is Changing the World and What We Eat*, London: Ebury Press, 2004.
8. Ibid., p. 200.
9. Ibid., pp. 197–202.
10. Kurlansky, *Cod*, p. 48.
11. Quoted in ibid., p. 55.
12. Ibid., pp. 43–4.
13. B. J. McCay and A. C. Finlayson, 'The Political Ecology of Crisis and Institutional Change: The Case of the Northern Cod', http://arcticcircle.uconn.edu/NatResources/cod/mckay.html
14. Glover, *The End of the Line*, p. 110.
15. R. B. Blake, 'Water Buoys the Nation: Fish and the Re-emergence of Canadian Nationalism', 1998, http://archiv.ub.uni-marburg.de/sum/90/sum90-6.html

16. Ibid., p. 3.
17. Ibid.
18. Ibid., p. 4.
19. Ibid., p. 5.
20. Ibid.
21. Ibid.
22. Ibid., p. 8.
23. Ibid.
24. Ibid.
25. Ibid., p. 2.
26. Ibid., p. 9.
27. Ibid.
28. *The United Nations Agreement for the Implementation of the Provisions of the United Nations Convention on the Law of the Sea of 10 December 1982 relating to the Conservation and Management of Straddling Fish Stocks and Highly Migratory Fish Stocks*, www.un.org/Depts/los/convention_agreements/convention_overview_fish_stocks
29. D. J. Bederman, 'CCAMLR in Crisis: A Case Study of Marine Management in the Southern Ocean', in H. N. Scheiber (ed.), *Law of the Sea: The Common Heritage and Emerging Challenges*, The Hague: Martinus Nijhoff Publishers, 2000, p. 174.
30. M. Lack and G. Sant, 'Patagonian Toothfish: Are Conservation and Trade Measures Working?', *TRAFFIC Bulletin*, Vol. 19, No. 1 (January 6, 2001).
31. Ibid., p. 6, http://www.ccamlr.org; M. Lack, 'Continuing CCAMLR's Fight Against IUU Fishing for Toothfish', WWF Australia and TRAFFIC International, 2008, www.wwf.org.au/publications/ccmlr
32. Seidl-Hohenveldern, 'Flags of Convenience', p. 301.
33. Lack and Sant, 'Patagonian Toothfish'.
34. www.colto.org
35. M. Peyron, 'Focus on Kerguelen, Hub of the Tooth-fish Saga', 2006, www.toothfishsaga
36. Viarsa l, COLTO, http://www.colto.org/Vessels/vess_ViarsaI.htm
37. See for instance, Peyron, 'Focus on Kerguelen'.
38. Ibid.
39. Ibid.
40. Patagonian toothfish (Chilean/Antarctic sea bass). Greenpeace, 2000, http://greenpeace.org/oceans/southernoceans/expedition20
41. J. Rubin, 'Piracy on the Cold Seas', *Audubon*, Vol. 100, No. 3 (1998), p. 22.
42. 'Patagonian Toothfish, *Seaweb*, 2001, http://www.seaweb.org/background/book/toothfish.html
43. G. Robertson and R. Gales, *Albatross Biology and Conservation*, Chipping Norton, N. S. W: Surrey Beatty & Sons, 1998, p. 18.
44. Save the Albatross, January 27, 2006, http://www.savethealbatross.net
45. Lack, 'Continuing CCAMLR's Fight', p. 1.
46. B. Davis, 'Coastal and Maritime Zone Planning and Management – Relationship Between Australia and the Southern Ocean/Antarctic', in M. Tsamenyi, S. Bateman and J. Delaney (eds), *Coastal and Maritime Zone Planning and Management: Transnational and Legal Considerations*, Wollongong, Australia: Centre for Maritime Policy, University of Wollongong, 1995, p. 118.

47. D. J. Bederman, 'CCAMLR in Crisis: A Case Study of Marine Management in the Southern Ocean', in Scheiber (ed.), *Law of the Sea*, p. 173.

48. 'General Introduction', CCAMLR, 2005, http://www.ccamlr.org/pu/e/gen-intro.htm

49. Davis, 'Coastal and Maritime Zone Planning and Management', p. 118.

50. Ibid.

51. Ibid., p. 119.

52. Bederman, 'CCAMLR in Crisis', p. 179.

53. Robertson and Gales, *Albatross Biology and Conservation*, Lack, 'Continuing CCAMLR's Fight'.

54. C. Roberts, *The Unnatural History of the Sea: The Past and Future of Humanity and Fishing*, London: Octopus Publishing Group Ltd, 2007.

55. Quoted in M. Tsamenyi and M. Herriman, 'Marine Biodiversity and the Law of the Sea', in C. M. Hawksley (ed.), *Science, Law and Policy for Management of the Marine Environment*, Wollongong, Australia: Centre for Maritime Policy, University of Wollongong, 2000, p. 76.

56. Fisheries Legislation Amendment Law 2007, Parliament of Australia.

57. Peyron, 'Focus on Kerguelen'.

58. Jacksonville News, November 13, 2006, http://www.jacksonville.com/apnews/stories/111306/DBLCA4NOO.shtml

59. G. Hardin, 'The Tragedy of the Commons', *Science*, Vol. 162 (1968), pp. 1243–8.

60. LOSC, Article 116.

61. Bederman, 'CCAMLR in Crisis', p. 176.

62. Quoted in J. M. Pureza, 'International Law and Ocean Governance: Audacity and Modesty', *International Law and Ocean Governance*, Vol. 8, No. 1 (1999), p. 74.

63. R. Baird, 'Political and Commercial Interests as Influences in the Development of the Doctrine of the Freedom of the High Seas', *Queensland University of Technology Law Journal*, Vol. 12 (1996), p. 290.

64. Bederman, 'CCAMLR in Crisis', p. 171.

65. Pureza, 'International Law', p. 74.

66. LOSC, Article 117.

67. Nellemann et al., *In Dead Water*, p. 5.

68. Ibid., p. 7.

69. Ibid., p. 36.

70. Ibid., p. 9.

71. L. Sneddon, V. A. Braithwaite and M. J. Gentle, 'Do Fishes have Nociceptors? Evidence for the Evolution of a Vertebrate Sensory System', *Proceedings of the Royal Society*, London B, Vol. 270 (2003), p. 1116.

72. Ibid., p. 1119.

73. Ibid.

74. Ibid., p. 1120.

75. J. Bentham, *Introduction to the Principles of Morals and Legislation*, New York: Hafner Publishing Co., 1948 (1789).

76. M. Aw, 'Cynanide and Dynamite Fishing – Who is Really Responsible?', 1996, http://141.84.51.10/riffe/infos/cyanide.html

77. G. Martin, 'The Depths of Destruction: Dynamite Fishing Ravages Philippines' Precious Coral Reefs', 2002, http://www.sfgate.com/cgi-bin/article.cgi?file=/chronicle.archive/2002/05/30.MN232485.DTL

78. M. J. Williams, 'Enmeshed: Australia and Southeast Asia's Fisheries', Lowy Institute for International Policy, Lowy Institute Paper 20, Sydney, 2007.
79. FAO, *The State of World Fisheries and Aquaculture 2006*, p. 7, http://www.fao.org/dorep/009/Ao699e/A0699EO4.htm#4.1.1
80. Worldwatch Institute, 'SOS for Fading Ocean Life', 2007, http://www.worldwatch.org/node/5360
81. Williams, 'Enmeshed'.
82. FAO, *The State of World Fisheries*, p. 7.
83. Roberts, *The Unnatural History of the Sea*, pp. 374–5.
84. Ibid., pp. 378–84.

6   CETACEANS AND THE SEA

1. Carl Sagan, *The Cosmic Connection*, Cambridge: Cambridge University Press, 2000.
2. G. Daws, '"Animal Liberation" as Crime: The Hawaii Dolphin Case', in Harlan B. Miller and William H. Williams (eds), *Ethics and Animals*, Clifton, New Jersey: Humana Press, 1983, pp. 361–73.
3. Richard Harrison and M. Bryden (eds), *Whales, Dolphins and Porpoises*, Sydney: Golden Press Pty Ltd, 1989.
4. Ibid.
5. Ibid., p. 98.
6. Denise Russell, 'Ethical Obligations to Cetaceans: A Case Study for Singer's Animal Ethics', in R. Younis (ed.), *On the Ethical Life*, Cambridge: Cambridge Scholarly Press, 2010.
7. Peter Singer, *Animal Liberation*, Oxford: Oxford University Press, 1990, p. 109.
8. See for example, Christopher Belshaw, *Environmental Philosophy: Reason, Nature and Human Concern,* Chesham, Bucks: Acumen Pub Ltd, 2001, pp. 118–19.
9. Anthony D'Amato and Sudhir K. Chopra, 'Whales: Their Emerging Right to Life', *American Journal of International Law*, Vol. 21 (1991), pp. 24–5.
10. IceNews, December 25, 2008, http://www.icenews.is/index.php/2008/12/25norway-lowers-2009
11. Planet Ark Environment News, May 1, 2008, http://www.planetark.com
12. Greenpeace Report, 'Japanese Government Gives in, Slashes Whale Quotas', November 13, 2008, http://www.greenpeace.org/international/news/whale-wars-quota-cut-131108
13. *Sydney Morning Herald*, November 21, 2008, p. 24.
14. ABC News, November 16, 2007.
15. IWC, 'Whale Population Estimates', November 3, 2008, pp. 1–2, http://www.iwcoffice.org/conservation/estimate.htm
16. Greenpeace International, 'Whale Meat Scandal', May 13, 2008, http://www.greenpeace.org/international/news/whale-meat-scandal-150408
17. The Marine and Coastal Community Network (MCCN), *NSW Bulletin*, August, 2006.
18. IWC, 58th Annual and Associated Meetings, St Kitts and Nevis, 2006, http://www.iwcoffice.org/meetings/meeting2006.htm
19. IWC, 'Aboriginal Subsistence Whaling', October 21, 2008, http://www.iwcoffice.org/conservation/aboriginal.htm

20. Harry N. Scheiber, 'Historical Memory, Cultural Claims, and Environmental Ethics: The Jurisprudence of Whaling Regulation', in Scheiber (ed.), *Law of the Sea*, p. 162.
21. B. Harnell, '"Secret" Dolphin Slaughter Defies Protests', *Japan Times*, No. 30, 2005, http://search.japantimes.co.jp/cgi-bin/fl20080330x1.html
22. *Sydney Morning Herald*, February 11, 2006.
23. Harnell, '"Secret" Dolphin Slaughter'.
24. IWC, 'Dolphin Slaughter', *Journal of Cetacean Research and Management*, Vol.7, Issue 2, (2005), p. 3.
25. Environment News Service, May, 2005, p. 6, http://www.ens-newswire.com/ens
26. N. Rau, *Sakhalin Success Good for Whales*, 2008, http://www.foe.co.uk
27. BBC News, September 5, 2005.
28. Environment News Service, May, 2005.
29. World Wildlife Fund for Nature Hong Kong 2008, 'Marine Pollution', Factsheet no. 30, http://www.wwf.org.hk
30. Defenders of Wildlife, *Climate Change: A Major New Threat to the World's Whales*, 2006, http://www.defenders.org/wildlife/new/marine/whales/climate.html
31. Ibid.
32. BBC News, September 5, 2005.
33. *Sydney Morning Herald*, October 27, 2005.
34. Ocean Futures Society, *Navy Sued Over Harm to Whales from Mid-Frequency Sonar*, November 1, 2005, http://www.oceanfutures.org/press/2005/pr_11_01_05.php
35. E. C. M. Parsons, Sarah J. Dolman, A. J. Wright, N. A. Rose and W. C. G. Burns, 'Navy Sonar and Cetaceans', *Marine Pollution Bulletin*, Vol. 56, No. 7 (2008), pp. 1248–57.
36. Defenders of Wildlife, *LFA Sonar: A Deadly Technology*, 2006.
37. Greenpeace, *The Dangers of Seismic Testing*, 2003, http://www.greenpeace.org/usa/news/the-dangers-of-seismic-testing
38. John May (ed.), *The Greenpeace Book of Dolphins*, Sydney: Random Century, 1990.
39. Harnell, '"Secret" Dolphin Slaughter'.
40. A. Cochrane and K. Callen, *Dolphins and Their Power to Heal*, London: Bloomsbury, 1992, p. 139.
41. Ibid., p. 132.
42. Quoted in ibid., p. 144.
43. LOSC, Article 65, p. 31.
44. Scheiber, 'Historical Memory', p. 132.
45. Ibid., pp. 135–6.
46. Ibid., p. 136.
47. Convention on International Trade in Endangered Species (CITES), *The CITES Appendices*, official documents, http://www.cites.org/eng/app/index-shtml
48. Steinar Andressen, 'The International Whaling Regime', in D. Vidas and W. Ostreng (eds), *Order for the Oceans at the Turn of the Century*, The Hague: Kluwer Law International, 1999, pp. 215–28.
49. CITES UK, 2006: News Conference of Parties, http://www.ukcites.gov.uk/news
50. Greenpeace, *Making Waves*, Issue 6, March 2006.
51. Greenmail, *Newsletter of the Greens*, NSW, March 2006.

7   SEA GYPSIES

1. C. Sather, *The Bajau Laut*, Kuala Lumpur: Oxford University Press, 1997.
2. J. Ivanoff, *The Moken*, Bangkok: White Lotus Co, 1997.
3. J. T. Thomson, 'Description of the Eastern Coast of Johore and Panang, with Adjacent Islands', *Journal of the Indian Archipelago and Eastern Asia*, Vol. 1 (1851), p. 140.
4. D. Sopher, *Sea Nomads: A Study Based on the Literature of the Maritime Boat People of Southeast Asia*, Singapore: National Museum Publication, 1997, p. 51.
5. C. Sather, 'Commodity Trade, Gift Exchange, and the History of Maritime Nomadism in Southeastern Sabah', *Nomadic Peoples*, Vol. 6 (2002), p. 24.
6. J. A. Crawfurd, *Descriptive Dictionary of the Indian Islands and Adjacent Countries*, London: Bradbury & Evans, 1856, p. 27.
7. T. Pires, *The Suma Oriental of Tomè Pires*, trans. and ed. A. Cortesâo, London: Hakluyt Society, 1944.
8. Sopher, *Sea Nomads*.
9. Pires, *The Suma Oriental*.
10. C. Sather, 'Sea Nomads and Rainforest Hunter-Gatherers: Foraging Adaptation in the Indo-Malaysian Archipelago', in P. Bellwood, J. J. Fox and D. Tryon, *The Austronesians: Historical and Comparative Perspectives*, Canberra: Department of Anthropology, ANU, 1995, p. 240; Sopher, *Sea Nomads*, pp. 50–1.
11. Sopher, *Sea Nomads*, pp. 229–68.
12. J. Ivanoff, *The Moken: Sea-Gypsies of the Andaman Sea: Post-war Chronicles*, Bangkok: White Lotus Co., 1997.
13. C. Chou, *Indonesian Sea Nomads: Money, Magic and Fear of the Orang Suku Laut*, London: Routledge Curzon, 2003.
14. Sopher, *Sea Nomads*, p. 263.
15. Ibid., pp. 122, 204.
16. Sather, 'Commodity Trade', pp. 20–44.
17. H. Nimmo, *The Sea People of Sulu*, San Francisco: Chandler Pub. Co., 1972, pp. 13, 16.
18. Chou, *Indonesian Sea Nomads*.
19. Ivanoff, *The Moken*, p. 1.
20. Quoted in Chou, *Indonesian Sea Nomads*, p. 33.
21. Ibid., p. 34.
22. T. Marett, 'Foreword' to W. G. White, *The Sea Gypsies of Malaya*, second edition, Bangkok: White Lotus Co., 1997, pp. 1–5.
23. See Sopher, *Sea Nomads*, p. 174 on the Moken, p. 130 on the Bajau, and p. 107 on the Orang Laut.
24. Chou, *Indonesian Sea Nomads*, pp. 34–5.
25. Ibid., p. 40.
26. Sopher, *Sea Nomads*, pp. 177–82.
27. A. R. Wallace, *The Malay Archipelago: The Land of the Orang Utan and the Bird of Paradise*, London: Macmillan and Co., 1893.
28. Sather, *The Bajau Laut*.
29. J. Verheijen, 'The Sama-Bajau Language in the Lesser Sunda Islands', *Pacific Linguistics*, Series D, Vol. 70, 1986, pp. 1–46.

30. Chou, *Indonesian Sea Nomads*.
31. White, *The Sea Gypsies of Malaya* (second edition).
32. J. Ivanoff, *Rings of the Coral: Moken Oral Literature*, Bangkok: White Lotus Press, 2001.
33. Sather, 'Commodity Trade', p. 25.
34. Sopher, *Sea Nomads*, p. 65; Sather, 'Commodity Trade', p. 35; W. G. White, *The Sea Gypsies of Malaya*, London: Seeley, Service & Co., 1922, pp. 57–8.
35. Sather, 'Commodity Trade', p. 36.
36. Pires, *The Suma Oriental*, p. 226.
37. White, *The Sea Gypsies of Malaya*, p. 144.
38. Ibid., pp. 144–5.
39. Nimmo, *The Sea People of Sulu*, p. 12.
40. Sather, 'Commodity Trade', p. 36.
41. Chou, *Indonesian Sea Nomads*, p. 29.
42. Ibid., p. 55.
43. Ibid., pp. 53–5.
44. Ibid., p. 65.
45. Ibid.
46. Ibid.
47. Ibid., p. 63.
48. Ivanoff, *The Moken*.
49. B. Bottignolo, *Celebrations with the Sun: An Overview of Religious Phenomena Among the Badjaos*, Manila: Ateneo de Manila University Press, 1995.
50. Chou, *Indonesian Sea Nomads*, p. 68.
51. Sopher, *Sea Nomads*, p. 135; Chou, *Indonesian Sea Nomads*, p. 68.
52. Sopher, *Sea Nomads*, pp. 121, 131.
53. Quoted in Ivanoff, *The Moken*, p. 73.
54. Sopher, *Sea Nomads*, p. 48; L. Lenart, 'Orang Suku Laut Identity: The Construction of Ethnic Realities', in G. Benjamin and C. Chou, *Tribal Communities in the Malay World: Historical, Cultural and Social Perspectives*, Leiden: Institute of Southeast Asian Studies and International Institute for Asian Studies, 2002, pp. 298–9.
55. Sopher, *Sea Nomads*, pp. 99–101, 307.
56. White, *The Sea Gypsies of Malaya*, pp. 273, 239; Nimmo, *The Sea People of Sulu*, pp. 15, 217; Sopher, *Sea Nomads*, p. 131.
57. Ivanoff, *The Moken*, pp. 127–8.
58. White, *The Sea Gypsies of Malaya*, p. 196.
59. Ibid., p. 197.
60. Ivanoff, *The Moken*, p. 74.
61. Sopher, *Sea Nomads*, p. 361.
62. Ibid., p. 308.
63. Ibid., p. 121.
64. Ibid., p. 131.
65. Australian Broadcasting Commission, *Burma's Gypsies of the Sea*, January 29, 2004.
66. Ivanoff, *The Moken*, p. 50.
67. White, *The Sea Gypsies of Malaya*, p. 42.
68. Nimmo, *The Sea People of Sulu*, p. 12.
69. Ibid.; Ivanoff, *The Moken*, p. 32.

70. A. Stuart, 'Forgotten Tsunami Survivors', *The Press*, Christchurch, January 11, 2005.
71. CBS News, March 20, 2005, www.cbsnews.com/stories/2005/03/18/60minutes/main681558.shtml
72. Ivanoff, *The Moken*.
73. Sopher, *Sea Nomads*, p. 93.
74. Ibid., p. 98.
75. Ibid., pp. 98, 123.
76. Ibid., p. 114.
77. Nimmo, *The Sea People of Sulu*, p. 15.
78. Ibid., pp. 15–16.
79. Ivanoff, *The Moken*, p. 20.
80. Chou, *Indonesian Sea Nomads*, p. 70.
81. White, *The Sea Gypsies of Malaya*, p. 87.
82. Sopher, *Sea Nomads*, p. 287.
83. White, *The Sea Gypsies of Malaya*, p. 87, 231.
84. Lenart, 'Orang Suku Laut Identity', p. 300.
85. Sather, *The Bajau Laut*, pp. 334–43.
86. Ivanoff, *The Moken*, p. 86.
87. Sopher, *Sea Nomads*, p. 147.
88. White, *The Sea Gypsies of Malaya*, p. 96.
89. Ivanoff, *The Moken*, p. 36, 47.
90. Personal observation.
91. Sopher, *Sea Nomads*, pp. 122, 152.
92. Ibid., p. 103.
93. Sopher, *Sea Nomads*, p. 122.
94. Lenart, 'Orang Suku Laut Identity', p. 299.
95. Ibid., pp. 299–300.
96. Ibid., p. 302.
97. 'Indonesia's Sea Gypsies Struggle to Survive', July 12, 2007, https://www.amazines.com/article_detail.cfm/391251?artcleid=391251
98. J. Josef and F. Munch, 'The Gentle People', 2004, http://www.trv.net/trv98/culture/gentlepeople
99. Chou, *Indonesian Sea Nomads*.
100. R. Ballint, *Troubled Waters: Borders, Boundaries and Possession in the Timor Sea*, Sydney: Allen & Unwin, 2005, p. 50.
101. Ibid., p. 61.
102. ABC News, October 18, 2007, www.abc.net.au/news/stories/2007/10/18/2062521.htm
103. N. Stacey, 'Crossing Borders: Implications of the Memorandum of Understanding on Bajo fishing activity in northern Australian waters', *Report for Environment Australia*, Canberra, 2001.
104. Commonwealth of Australia, 'Ashmore Reef National Nature Reserve and Carter Island Marine Reserve Management Plans', *Report for Environment Australia*, Canberra, 2001.
105. Ballint, *Troubled Waters*, p. 101.
106. Ibid., p. 99.
107. Ibid., p. 102.
108. Quoted in ibid., p. 96.
109. Quoted in ibid., p. 3.

8 INDIGENOUS SEA CLAIMS

1. In Anon, *Yirrkala*, Buku-Larrngay Mulka Centre, 1999.
2. Roslynn D. Haynes, *Seeking the Centre: The Australian Desert in Literature, Art and Film*, Cambridge: Cambridge University Press, 1998, p. 12.
3. Nin Tomas and Kerensa Johnston, 'Ask That Taniwha: Who Owns the Foreshore and Seabed of Aotearoa?', September 12, 2003.
4. Constitution of the Haida Nation, 2006, p. 5, www.haidanation.ca
5. Gary D. Meyers, Malcolm O'Dell, Guy Wright and Simone C. Muller, *A Sea Change in Land Rights Law: The Extension of Native Title to Australia's Offshore Areas*, Canberra: Australian Institute of Aboriginal and Torres Strait Islander Studies, 1996, p. iii.
6. J. Bradley, '"We Always Look North": Yanyuwa Identity and the Maritime Environment', *Oceania Monograph*, No. 48 (1998), pp. 125–41.
7. Paul Moon and Peter Biggs, *The Treaty and its Times*, Auckland: Resource Books, 2004, p. 201.
8. Ibid., p. 464.
9. Ibid., p. 391.
10. A. Searcy, *In Australian Tropics*, London: George Robertson and Co., 1909, pp. 84, 174.
11. Kristen M. Fletcher and Stacy Prewitt, *Tribal Fishing Rights Take Precedent in Ninth Circuit*, 1998, http://www.olemiss.edu/orgs/SGLC/MS-AL/tribal.htm
12. Constitution of the Haida Nation, pp. 5–6.
13. M. Hume, 'Haida sue for Queen Charlottes', *Haida Laas: Journal of the Haida Nation* (April 2002), p. 3.
14. Quoted from K. Petersen, 'The Struggle for Haida Gwaii', *The Dominion*, November 6, 2004, p. 2, http://www.dominionpaper.ca/original_peoples/2004/11/06/the_strugg.html
15. G. Baker, 'NaiKun: Shell or NaiKun', *Haida Laas: Journal of the Haida Nation* (August 2009), pp. 1–2.
16. Tomas and Johnston, 'Ask That Taniwha', p. 7.
17. Maria Bargh, 'Changing the Game Plan: The Foreshore and Seabed Act and Constitutional Change', *New Zealand Journal of Social Sciences Online*, Vol. 1 (2006), p. 14.
18. Quoted in ibid., p. 16.
19. High Court of Australia, Decision of the High Court of Australia, October 11, 2001, updated January 29, 2002, *The Commonwealth v. Yarmirr*; *Yarmirr v. Northern Territory*.
20. Federal Court of Australia, *The Commonwealth of Australia v. Yarmirr and Ors*; *Yarmirr and Ors v. the Northern Territory of Australia and Ors*, Summary of Judgment by Justice Olney, 1997.
21. High Court of Australia, Decision of October 11, 2001.
22. Denise Russell, 'Aboriginal-Makassan Interactions in the Eighteenth and Nineteenth Centuries in Northern Australia and Contemporary Sea Rights Claims', *Australian Aboriginal Studies*, No. 1 (2004), pp. 3–17.
23. High Court of Australia, Decision of October 11, 2001. p. 72.
24. Ibid., p. 74.
25. Federal Court of Australia, *Gumana v. Northern Territory* (Justice Selway), 2005.

26. Federal Court of Australia, *Gumana v. Northern Territory* (Justice Mansfield), 2005.
27. Federal Court of Australia, *Gumana v. Northern Territory*, 2007, p. 3.
28. High Court of Australia, *Northern Territory of Australia v. Arnhem Land Aboriginal Land Trust*, 2008.
29. Ibid., pp. 11–12.
30. Ibid., p. 14.
31. Ibid., p. 17.
32. Planet Ark News, October 4, 2007, phttp://www.planetark.com/dailynewsstory. clm/newsid/44654/story.htm

9   PROTECTION OF THE OCEANS

1. Geouffre de la Pradelle, 'Le droit de l'État sur la mer territoriale', *Revue Générale de Droit International Public* (1898), p. 282.
2. See for instance C. Perrings and M. Gadgil, 'Conserving Biodiversity: Reconciling Local and Global Public Benefits', in I. Kaul, P. Conceição, K. Le Goulven and R. Mendoza (eds), *Providing Global Public Goods: Managing Globalisation*, New York: United Nations Development Programme, 2003, pp. 532–55; D. A. Posey, 'Fragmenting Cosmic Connections: Converting Nature Into Commodity', in S. Vertovec and D. Posey (eds), *Globalisation, Globalism, Environments, and Environmentalism: Consciousness of Connections*, Oxford: Oxford University Press, 2003, pp. 123–40.
3. Convention on Biological Diversity, Article 8, http://www.biodiv.org/ convention/articles.asp?1g=0&a=cbd-08
4. National Marine Sanctuaries, *Native Cultures and the Maritime Heritage Program*, December 28, 2005, http://sanctuaries.noaa.gov/maritime/cultures. html
5. Marcia Langton, Zane Ma Rhea and Lisa Palmer, 'Community-Oriented Protected Areas for Indigenous Peoples and Local Communities', *Journal of Political Ecology*, Vol. 123 (2005), p. 29.
6. Ibid.
7. Ibid., p. 32.
8. Ibid., p. 36.
9. *Sydney Morning Herald*, November 24, 2008, p. 6.
10. Glover, *The End of the Line*, p. 44.
11. F. Montaigne, 'The Global Fish Crisis: Still Waters', *National Geographic* (April 2007), p. 1, http://environment.nationalgeographic.com/environment/ habitats/global-fish-crisis.html
12. Panda News, October 12, 2008, http://www.panda.org/news_facts/newsroom/ press_releases/index.cfm?uNewsID=147801
13. 'New Bluefin Catch Rule Riles Scientists', November 2008, http://features. csmonitor.com/environment/2008/11/28/new-bluefin-catch-rule-riles-scientists
14. Planet Ark News, September 26, 2007, http://www.planetark.com/ dailynewsstory.cfm/newid/44506/story.htm
15. Ibid.
16. Kaul et al., *Providing Global Public Goods*, p. xiii.
17. IUCN: The World Conservation Union, 'New Ocean Threats Underline Need For Urgent Action to Protect the High Seas', October, 2007, http://www.iucn. org/en/news/archive/2007/10/24_pr_marine.htm

18. IPCC, Intergovernmental Panel on Climate Change, p. 1, http://www.ipcc.ch/about/index.htm
19. The International Seabed Authority, Convention and Agreement, www.isa.org.jm/en/home
20. Ibid.
21. *Sydney Morning Herald*, March 30, 2006, p. 8.
22. *Sydney Morning Herald*, January 7, 2009, p. 9.
23. Marine Stewardship Council, 2008, p. 1, http://www.msc.org/healthy-oceans

Index

Abbas, Abul 78–80
Aboriginal and Torres Strait Islanders 4
  beliefs about ownership 137–40
  Blue Mud Bay case 5, 142, 144–8, 152, 153
  Croker Island case 142–5, 147
  Mabo judgment 139–40
  sacred sea places 48
Achille Lauro (ship) 77–83
Africa 91, 102
  Algiers 8–12
  Ethiopia 68, 71
  North African pirates 7–12
  Somalia *see* Somalia
  South Africa 92, 93
  West Africa 155
  see also Gulf of Aden, Somali piracy
Alaska 14
  climate change 148–9
  Exxon Valdez (ship) 35–7, 39
  oil and gas 29, 37, 40, 42
  Sarah Palin 42
  sea rights claims 3, 33, 140–1
  whaling 110–11
Algiers 8–12
Alondra Rainbow (ship) 60–3
animal ethics 4, 100–5, 107–16, 118, 120
Antarctica 118, 119
  Antarctic Treaty 32, 33, 153
  exploration 45
  ice sheets 27, 32
  Southern Ocean Sanctuary 109–10
archaeology 49–52, 53, 58, 59
  corrupted 51–5
Arctic, the 113, 148–9
  climate change 40–2
  exploration/exploitation 1, 6, 33–40, 45–6, 153
Arctic Council 35–40

Arctic Ocean 1, 6, 29, 32, 33–42, 45, 163
Argentina 32, 93
armed robbery against ships 63–4, 66
Atlantic Ocean, 156
  fishing 86, 88–91, 156
  ownership claims 7, 11, 19, 26
  piracy 10
  shipwrecks 3
  whales 112
Attenborough, Sir David 94–5
Australia 72, 131
  Antarctica 32, 33
  fishing 26, 92–4, 97–9, 152, 163–4
  indigenous peoples *see* Aboriginal and Torres Strait Islanders
  sea gypsies 134–6
  sea refugees 27
  underwater cultural heritage 48, 53–5
  whale watching 116, 119
  whaling 109–10, 113, 114, 119
Australian High Court decisions *see* Blue Mud Bay case; Croker Island case; Mabo judgment
Australian Maritime Information Zone 24
Austria 78

Bahamas, the 114, 115
Ballint, R. 134, 135
Bangladesh 64
Barbarossa (pirate) 8, 9
baselines 32
Bederman D.J. 95, 96
Belize 82, 92
Bentham, Jeremy 102, 107
Blue Mud Bay case 5, 142, 144–8, 152, 153
Bonino, Emma 89, 90
Borg, Joe 34
Bottingnolo, Bruno 126–7
Braudel, Fernand 10

Brazil 75
Britain 125, 139, 141–2
  fishing 4, 25, 84–7, 97
  oil and gas 33
  piracy 68, 73, 76, 78, 79
  sea claims 3, 17, 19, 22, 32
  underwater cultural heritage 48, 50,
    57–9
Burma (Myanmar) 70, 122, 125, 128,
  129, 134
Bush, George W. 30, 161

Caboto, Giovanni 87, 90
Canada 3, 4, 41–2, 58, 72, 149
  Arctic Ocean claims 33, 34–5, 40
  fishing 86–90, 98, 99, 155, 156,
    162
Caribbean, the 75
Carson, Rachel 2
Cassese, Antonio 77–82
CCAMLR (Commission for the
  Conservation of Antarctic Marine
  Living Resources) 92, 93, 95–7,
  156, 162
Chile 93, 97
  sea gypsies 121
China 51–3, 72, 102, 127, 134
Chou, Cynthia 122, 126–8, 134
Christian pirates 10
CITES (Convention on International
  Trade in Endangered Species)
  118, 120
climate change 2, 5, 26, 103, 120, 132,
  146, 153, 159
  Arctic 38, 40–2
  coral 27–8, 44, 100, 133, 148
  fish 10
  indigenous sea claims, impacted by
    148–9
  Intergovernmental Panel on 40, 159
  low-lying islands 26–8, 161
  Oceans Council and 160
  permafrost melts 37, 40–1
  rising sea levels 26–8, 41
  runaway greenhouse effects 43, 44
  sea gypsies 132
  storminess 148
  undersea mining 2, 29, 153, 160
  whales 113
  see also greenhouse gases

closed seas 6, 31
  Grotius' views on 19–20
  Law of the Sea Convention 24
  Selden's views on 18–21
  Venice and Adriatic Sea 6–7
COLTO (Coalition of Legal Toothfish
  Operators) 92–3
Combés, F. 128
continental shelf 22, 23, 31–3, 39, 90,
  140
coral 161
  bleaching 27–8, 44, 100, 133
  fish 100, 130, 132–3
  ocean acidification 43, 133, 148
  sea temperature 44, 148
Cousteau, Jacques 116
Crete 102
Croker Island case 142–5, 147
cruise ships (actual or attempted
    hijackings) 68
  Achille Lauro 78–82
  Melody 68
  Seabourn Spirit 70, 77
cultural threats to sea gypsies 132–6
Cyprus 92

Dampier, William 13
Denmark 3, 72
  Arctic Ocean claims 33–5
  underwater cultural heritage 48, 53
  see also Greenland
Dillon, Dana 66, 67
dolphins 106, 113, 129, 157, 163
  bottlenose 112
  captive 115–16, 119
  ethical concerns 101, 105, 111, 118,
    120
  Greenpeace and 118–19
  slaughter 111, 112, 115, 157
  vaquita 106, 112
  see also whales
Dominican Republic 82
Drake, Francis 11, 13
Dromgoole, Sarah 50
Dutch East India Company 11, 21
  shipwrecks 54–5

Earle, Sylvia 25
East Indies 10, 11, 17
Ecuador 156

Egypt 70, 77–80
Elizabeth I, Queen 13
Estai, (ship) 89–90
ethics
  animal ethics 4, 100–5, 107–16,
    118, 120
  intergenerational *see* future
    generations
Europe 21, 40, 48, 156
European Union (EU) 34–5, 88
  Arctic research 1
  Arctic resources 3
  fishing wars 4, 73, 88–90
  piracy 72
Exclusive Economic Zone (EEZ) 27,
    98–9, 136
  in Australia 24, 99, 135
  defined 22–4
  fishing 88–93, 97, 103–4
  in new proposal on ocean
    governance 153–7, 163
  piracy 62, 65–6, 82
  underwater non-living resources 26
  see also high seas; territorial sea
exploitation
  Arctic 1, 6, 33–40, 45–6, 153
  Law of the Sea and 99–101
  see also oil and gas exploitation;
    over-exploitation of fish

Falkland Islands 93
FAO (Food and Agricultural
    Organisation) 25, 103
Faroe Islands 86
Finland 35
fish 26, 149, 161, 162
  in Arctic 38, 51
  coral reef decline 28, 132–3
  historical views of 16, 18, 22
  marine pollution and 83, 102
  ocean acidification, impacts of 43,
    100
  pain 101
  white 18, 25
  see also fishing, over-exploitation of
    fish
fish (or fishery) piracy 92–100
fishing 84–104, 144, 155
  in Atlantic Ocean 86, 88–91, 156
  for Atlantic salmon 86
  climate change effects on 100
  for cod 84–8, 98, 113, 162
  ethical concerns 100–5, 161
  for herring 85
  illegal *see* illegal fishing practices
  long-line 96
  marine pollution and 86, 88–100,
    102–4
  moratoriums 25, 87, 88, 92, 156,
    161
  ocean acidification effects on 100
  for Patagonian toothfish 92–8, 156
  quotas *see* quotas
  for sharks 134
  for trout 101
  for tuna 110, 156, 164
  for turbot 88–92, 99
  see also fish; over-exploitation of
    fish.
fishing fines 97
fishing licences 26, 86, 95, 146, 147,
    152, 161, 162
fishing nets 119, 133, 162
fishing observers 162
fishing reserves *see* marine sanctuaries
fishing rights 157
  conflicts over 67, 150
  in Exclusive Economic Zones 23
  indigenous peoples and 143–8, 152
fishing subsidies 161
fishing wars
  cod wars 84–8
  over Patagonian toothfish 92–100
  turbot wars 88–92
  war on fish 4, 100–4
flags of convenience 56–7, 81–2, 88,
    92, 95, 156, 159
fossil fuels in seabeds, 29, 32–3, 35;
    see also oil and gas exploitation
France 76
  fishing 84, 87, 92–3, 163–4
  Kerguelen Island 32, 92, 97, 99
  piracy patrols 72–4
  underwater cultural heritage, 48, 53,
    58
freedom of the seas 120, 150
  challenges to 2, 5, 6–14, 17–23, 30,
    117
  Grotius's arguments for 14–17
  indigenous peoples and 142–8

Law of the Sea and 23–6
limitations on 21–3
Mare nullius 140
see also fishing wars; new proposal
  for ocean governance
future generations 141, 164
  the Arctic environment 40, 46
  CITES obligations to 118
  freedom of the seas doctrine and 25
  marine pollution 153
  species survival 102, 103, 157, 160
  underwater cultural heritage 55–6,
    158
  whale numbers 108

Gentili, Alberico 13, 65
Germany 55, 72, 76, 78, 79, 160
  fishing 84–6
Greece 72, 74, 115
  underwater cultural heritage 49
greenhouse gases 36, 42
  carbon dioxide 36–8, 41, 43–4, 153,
    159, 163
  methane 36–7, 41
Greenland 3, 34, 35, 40, 149
  whaling 110
Greenpeace 156
  dolphins, protection of 118–19
  fish, protection of 92, 156
  whales, protection of 110
Grotius, Hugo 3, 90, 142–5, 150, 152
  arguments for freedom of the seas
    14–17, 140, 144, 146
  piracy problem 7
  replies to arguments 17–21
Gulf of Aden 61, 64, 67, 71, 72, 75, 76

Hardin, G. 98
Harper, Stephen 35
Heard and McDonald Islands 23, 92
Herman, Louis 105
high seas
  fishing 88–100
  freedom of 2, 6, 23, 26
  hijacking of ships on 25
  piracy on see under piracy
  seabed of 30–1
  terminology 154, 156, 157, 163
  underwater cultural heritage 56, 58,
    59

see also Exclusive Economic Zone;
  fishing wars; hijacked ships;
  international waters; seabed;
  territorial seas
hijacked ships 154–5
  Achille Lauro 77–83
  disappearance of 25, 60–3
  number of 65
  on high seas 25
  by Somali pirates 62, 64, 67–70, 72,
    74, 76
Hong Kong 70

ICCAT (International Commission for
  the Conservation of Atlantic
  Tunas) 156, 162
Iceland 4, 35
  fishing 84–6, 88, 90–1, 98, 155, 162
  whaling 108
illegal fishing practices 56, 84–9, 110,
  133–5, 152, 155–6, 163–4
IMB (International Maritime Bureau)
  63–69, 71, 74, 75
IMO (International Maritime
  Organization) 63, 65, 66, 69, 75,
  160
India 63, 64, 72, 74
Indian Ocean 29, 72
indigenous peoples 45, 148, 150–3
  in Australia see Aboriginal and
    Torres Strait Islanders
  fishing rights 143–8, 152
  freedom of the seas, problems for
    142–8
  Haida nation 138, 141
  Inuit 40, 46, 149
  Kaiganii 141
  Maori 137–9, 141–2
  mining objections 140–1, 149
  oil and gas 39, 40
  sea claims 5, 137–49, 163
  Shishmaref community 148–9
  whaling 110–11, 117, 157
Indonesia 164
  fishers 134–5
  navigational rights 24
  piracy 60–4, 66–7, 70
  Sulawesi 121, 127, 131
  tsunami (2004) 133
  see also Straits of Malacca

Intergovernmental Panel on the Oceans (proposed) 159
International Commission for the Conservation of Atlantic Tunas *see* ICCAT
International Court of Justice 91
International Fishing Authority (proposed) 160–3
International Maritime Bureau *see* IMB
International Maritime Organization *see* IMO
International Seabed Authority *see* ISA
International Union for the Conservation of Nature *see* IUCN
international waters
    redefined 84, 153–8
    use of term 23, 66, 71, 79
    see also new proposal for ocean governance
International Whaling Commission *see* IWC
Iran 72
Ireland 53
ISA (International Seabed Authority) 29–31, 56, 160–2
Israel 78
    attack on USS *Liberty* 82
Italy 77–81, 156, 164
IUCN (International Union for the Conservation of Nature) 120, 152, 156, 159, 164
Ivanoff, J. 121, 122, 125, 129, 130
IWC (International Whaling Commission) 111–20, 157
    Japanese whaling 108–10
    Law of the Sea Convention 117, 120

Jamaica 53
Japan 60, 61, 93, 128, 156, 164
    International Whaling Commission and 108–10
    whaling 108–11, 118, 119, 157
Juno (shipwreck) 54

Kaul, I. 158
Kenya 73
Kirby, Michael 143–5
Kiribati 27, 161
krill 43, 87, 96, 113

Kurlansky, Mark 84, 87

La Capitana Jesus Maria (shipwreck) 54
La Galga (shipwreck) 54
Langton, Marcia 151–2
Law of the Sea Convention *see* LOSC
Lenart, L. 131
Liberia 92
Libya 78
Lomonosov mountain chain (Arctic) 33
Lopez, Barry 45–6
LOSC (Law of the Sea Convention) 154
    alternative proposals *see* new proposal for ocean governance; Oceans Council
    exploitation, promoted by 99–101
    Grotius' principles in 23–6, 143
    international conflict and 34, 35, 55, 120, 153
    International Seabed Authority 29–31, 56, 160–2
    piracy 3, 7–13, 21, 61–8, 73, 77, 82, 155
    problems with 2, 34, 52, 57, 82, 91, 96–100, 120, 136
    sea zones 22–3, 98
    signatories 90
    underwater cultural heritage 52–3, 55, 56, 58, 59, 158
    US failure to ratify 29–31, 58
    whales 117, 120
    see also IWC
low-lying islands 26–8, 161

Mabo judgment 139–40
Macassans, the 143
Malaysia 164
    piracy 61, 62, 64, 70, 82
    piracy counter-measures 66, 72, 82, 164
    sea gypsies and 125–8, 131, 134
    see also Straits of Malacca
Maldives 27
marine commons 98, 99
marine environmental management 25
    in Antarctica 32–3, 92–100, 110
    in the Arctic 35

climate change 44
Fisheries Management Act 97
fishing 86, 88–100, 102–4
flags of convenience 57
in Iceland 86
Law of the Sea 23
in new proposal for ocean
    governance 26, 150–64
whales and 109, 110, 112, 114,
    115, 117–20
marine pollution
chemical contaminants 112–3
Clean-up the World campaign 162
fish 83, 102
flags of convenience 56–7, 159
noise 113–6, 120
ocean circulation 100
from oil and gas activities 35–40,
    112, 149
poisons 133
ship-based 159
waste, dumping of 69, 159
marine sanctuaries 151
for fish 96, 98–9, 104, 135, 145,
    161
for whales 109, 110, 114, 117–19
Marine Stewardship Council 162, 164
maritime security 24
Mediterranean Sea 156
piracy 7, 8, 10, 12
shipwrecks 49
Middleton, Roger 68
Moors 7–9
mother ships 25, 69, 73, 74

NAFO (Northwest Atlantic Fisheries
    Organization) 88–91, 162
Nansen, Fridtjof 45–6
NATO (North Atlantic Treaty
    Organization) 72, 73, 79, 85, 86,
    115
navigation 16, 17, 22
licences 26
rights 23–5, 62, 140–3, 145, 147,
    148, 150, 154
by sea gypsies 129
Netherlands, the 71, 84
counter-piracy measures 72–3
historical sea voyages 7, 11, 13–15,
    17, 19

Netherlands Antilles 156
underwater cultural heritage 52,
    54–5
new proposal for ocean governance 5,
    55, 82, 99–100, 120
for coastal waters 150–3
for sea beyond 12 nautical miles 26,
    59, 66, 82–3, 91–2, 104,
    153–64
see also freedom of the seas; ocean
    governance
New Zealand 24, 32, 93
Maori 4–5, 137–42
piracy 68
sea refugees 27, 32
Treaty of Waitangi 5, 139, 141
Nigeria 64
Nimmo, H. 126
North African pirates 7–12
links to Ottoman Empire 9
North Atlantic Treaty Organization see
    NATO
North Pole 34, 45
North Sea fishing 84
Northwest Atlantic Fisheries
    Organization see NAFO
Northwest Passage 149
Norway 3, 26, 33–5, 91
whaling 108, 111, 118
Nuestra Señora de Atocha (shipwreck)
    54
Nuestra Señora de la Mercedes
    (shipwreck) 57–8

Obama, Barack 31
ocean acidification 2, 28, 43–4
effects on fish 43, 100
see also greenhouse gases
Ocean Alliance 112
Oceans Council (proposed) 158–61
ocean governance 25, 84, 89
fishing 90–2, 95, 100
piracy 65, 67, 73–5, 82–3
underwater cultural heritage 55–9
whales 105, 114, 117–18, 120
see also Law of the Sea Convention;
    new proposal for ocean
    governance
Oceania Project 119
Odyssey (savage company) 57–9

oil and gas exploitation
  in Antarctica 32–3
  in the Arctic 1, 34–42, 163
  degradation of habitat 35–40, 112,
    115, 154
  International Seabed Authority 30,
    160
  Norway 26
  indigenous peoples' objections to
    39, 40, 140–1, 149
  seabed of the continental shelves 23,
    31
  seabed of the Exclusive Economic
    Zones 29–30
O'Malley, Grace 13
over-exploitation of fish
  illegal see illegal fishing practices
  moratoriums, need for 25, 87, 88,
    92, 156, 161
  in Southeast Asia 133–5
  war on fish 4, 100–4
  see also fishing
ownership claims on sea 25, 122
  continental shelf 31
  high seas 98, 103, 117
  historical 21–3, 87–8
  by indigenous peoples 5, 137–49,
    163
  in new proposal for ocean
    governance 103–4, 150–64
  territorial sea 26, 104

Pacific Ocean 29, 105, 112, 141, 156,
  161
Pakistan 72
Palin, Sarah 42
Panama 56, 60, 92, 93, 156
Papal Bulls 11, 17, 21
Papua New Guinea 75
Pardo, Arvid 29
permafrost 37, 40–1
Philippines 102, 122, 133
piracy 89, 90, 103–4, 122
  actual and attempted attacks 64–5
  Alondra Rainbow (ship) 60–3
  against the British 10
  attacks on private boats 75–6
  defined 62–3
  fish (or fishery) piracy 92–100, 156

on the high seas 61–6, 72, 76, 81,
    82, 103–4, 117, 155
  law of the sea and 3, 7–13, 21,
    61–8, 73, 77, 82, 155
  in new proposal for ocean
    governance 84, 154, 159, 160,
    163
  prosecutions 25, 63, 65–7, 79–83,
    154, 160
  ransom demands 70–5, 77
  reasons for 67–70
  as term of condemnation 89, 90
  treasure salvors and 51
  see also Somali piracy; terrorism on
    the sea
Pires, Tomè 125
plankton 43, 105, 113
PLO (Palestine Liberation
  Organisation) 77–80
polar bears 39, 42, 45–6, 149
Portugal
  counter-piracy measures 74
  fishing 88–91
  ownership claims on Atlantic Ocean
    7, 11, 15–19, 26
  underwater cultural heritage 53, 56
privateers 11–13
Pureza, J.M. 99

quotas
  for fish 86, 88, 89, 91, 110, 146,
    156, 161
  for whales 108, 110, 117

Reagan, Ronald 30, 79
Red Sea 61, 64
regional agreements
  fishing 23, 140, 155
  piracy 67, 82
  see also CCAMLR, ICCAT
right of innocent passage 22–4, 68,
  143, 147
rights see fishing rights; navigation
  rights; right of innocent passage;
  sovereign rights
Roberts, C. 96, 104
Rudd, Kevin 148
Russia
  anti-piracy moves 72
  Arctic claims 1, 3, 33–5

illegal fishing 92–4
indigenous peoples 149
mining accidents 37, 39
permafrost and climate change
    40–1
Siberia 33, 41, 110
whales 112, 115

salvage laws 49–53, 56–9, 158
Sandalow, David 30, 31
Sanho (ship) 60–1
Sather, C. 121, 122, 125, 126
sea 21, 43 *see also* closed seas; freedom
    of the seas; ownership claims on
    sea
seabed 23, 29–31, 52, 141–3, 153, 158
    see also ISA
seabirds 38–9, 46
    albatross 94–6
sea boundaries 67, 134–6, 144, 155,
    157, 159, 163
sea claims *see* ownership claims on sea
sea gypsies 150
    Bajau Laut, 121, 122, 124–6, 128,
        130–1, 134, 135
    boats 123–6, 128–31, 135
    cultural threats to 132–6
    fishing practices 124, 126, 130, 131,
        133
    languages 121, 122, 125, 128,
        130–1, 134
    Moken 121, 122–31, 133, 134, 152
    Orang Laut 122, 124, 125–8,
        130–4
    religions 12–17, 131–2, 134
sea levels 26–8, 42, 159
sea refugees 27–8
Sea Shepherd Conservation Society
    109, 119
sea temperatures 44, 100
sea zones 22–3, 98 *see also* Exclusive
    Economic Zone; high seas;
    international waters; territorial
    sea
seals 38, 96, 113, 149
Searcy, Alfred 139
Selden, John 18–21
Seychelles, the 27, 61, 73, 76
Shakespeare, William 47, 87
sharks 130, 134

shipping 24, 65–9, 72, 75, 149, 154,
    159
    noise 113
    registration 24, 56–7, 159
shipwrecks 47–59, 160, 163
    Spanish 54, 57–8
Singapore 66, 69, 82, 92, 122
Singer, Peter 107
Sirius Star (ship) 67, 69
Soloman Islands 75
Somalia
    foreign fishers 23, 66, 69, 91,
        155–6
    piracy *see* Somali piracy
Somali piracy 164
    hijacked ships 62, 64, 67–70, 72,
        74, 76
    links to inshore conditions 69, 75
    number of attacks 61, 64, 71
    prosecution dilemmas 72–5, 154
    rise in 70–5
    Sirius Star (ship) 67, 69
    territorial waters 25
    treasure salvors 51
    UN 25, 68, 70–2, 82, 154
    weapons 67–8
    on yachts 76
Sopher, D. 122, 128, 130, 131
Southeast Asia 67, 69, 91, 102, 103,
    152, 155–6
    sea gypsies of 121–36, 150
sovereign rights 22, 23, 25, 26, 29, 30,
    56, 91
    out to 350 nautical miles 31, 35
Southern Ocean 99, 163, 164
    fish (or fishery) piracy 92–8, 156
    fossil fuels in seabeds 29, 32–3, 35
    Heard and McDonald Islands 23, 92
    krill 113
    ocean acidification 43–4
    Southern Ocean Sanctuary 109–10
    whales 119
Spain
    fishing 87–90, 92, 93, 155, 156, 164
    ownership claims on the Atlantic
        Ocean 11, 17–18, 26
    piracy 7–8, 72, 73, 82
    shipwrecks 54, 57–8
    underwater cultural heritage 53,
        56–8

Spice Islands 21
storm surges 28
Straddling Stocks Agreement 90, 91, 93, 98–9, 103, 155, 156
Straits of Malacca 24, 60–2, 64, 66, 67, 69, 82, 164
submerged villages 48
subsoil 23, 31
Sweden 35, 40, 76
Syria 78

Taiwan 133, 134
territorial sea
    in Australia 144
    in Britain 58
    climate change 26–7
    historical demarcations of 22, 23, 85
    Law of the Sea 23, 26, 61–2, 65
    in new proposal for ocean governance 66, 91–2, 104, 150–7
    piracy 25, 72, 81
    right of innocent passage in 22–4, 68, 143, 147
    in Somalia 25, 72
    underwater cultural heritage 50, 52, 66
    see also Exclusive Economic Zones; high seas
terrorism on the sea 76–82
    fear of 24
    terrorism defined 77
Thailand 61, 70, 74, 122, 129, 133, 152
Thomson, J.T. 121
Titanic (ship) 58
Tobin, Brian 84, 88–90
Torres Strait Islanders see Aboriginal and Torres Strait Islanders
treasure salvors 49–52, 57–9
Treaty of Waitangi 5, 139, 141
Truman, Harry 22
Tunisia 53, 78–80
Turkey 49, 50, 55
Tuvalu 27

Ukraine 93, 97
underwater cultural heritage 49–54, 84, 160, 163
    convention about 56–8, 157–8

defined 47–8
    in high seas 56, 58, 59
    in Law of the Sea Convention 52–3, 55, 56, 58, 59, 158
    shipwrecks 47–59, 160, 163
    see also salvage laws
UN (United Nations)
    anti-piracy Convention and Protocol 81–3, 160
    charter on use of force in international relations 74, 80
    continental shelf claims 32, 33
    Convention on Biological Diversity 151
    Development Program Report 27
    drift net fishing resolution 119
    Environmental Program 100, 159, 162, 164
    FAO (Food and Agriculture Organisation) 25, 103
    future role 26, 28, 158, 163, 164
    Intergovernmental Panel on Climate Change 159
    IMO see IMO
    Law of the Sea Convention see LOSC
    NAFO see NAFO
    on racial discrimination 142
    Somali pirates and 25, 68, 70–2, 82, 154
    Straddling Stocks Agreement see Straddling Stocks Agreement
    UNESCO Conventions 55–9
    underwater cultural heritage 56, 59
Uruguay 93
US (United States) 3, 22, 151, 156, 161, 164
    illegal fishing 97
    indigenous sea rights claims 138, 141
    Law of the Sea Convention, failure to ratify 29–31, 58
    Native Cultures and Maritime Heritage Program 151
    oil and gas mining 29–30, 34, 35, 40
    piracy 71–4
    terrorism on the sea 77–80, 82
    underwater cultural heritage 53, 54, 57–9

US Fish and Wildlife Service 109
    whales 112, 114, 115, 119
    see also Alaska
USS *Liberty* 82

Venezuela 75, 76, 156
Verheijen, J. 125
Viarsa I (ship) 93
Vietnam 64, 102

Wallace, Alfred Russel 25
Welwod, William 17–18, 25
whales 4, 43, 109, 163
    beluga 106, 112, 113
    blue 105–6, 113
    bowhead 106
    climate change 113
    commercial fishing and 112
    diet 106
    ethical concerns 101, 105, 107–11,
        118, 120
    fin 106
    habitat degradation 112–13
    humpback 106
    migration 106–7, 111, 119
    minke 106
    narwhal 106, 113
    orca 106, 115
    Pacific Gray 38, 42, 106, 111–12
    pilot 106, 114

protection agencies 117–20, 164
    right 106, 112, 119
    sanctuaries *see* marine sanctuaries
    scientific permits for 108–9, 117,
        157
    seismic testing and 115, 157
    sonar technologies and 114–15,
        120, 157
    sperm 94, 106, 115
    strandings 114, 115
    Western grey 106, 115, 120
    whale watching 116–17
    see also dolphins
whaling
    bowhead whales 111
    fin whales 108, 109
    humpback whales 109, 113,
        116–17, 119
    minke whales 108, 109, 113, 118
    moratorium 101, 108, 111, 117,
        118, 157
    quotas *see* quotas
    see also IWC
White, W.G. 125, 126, 128–31
World Conservation Union 112, 114
World Wildlife Fund 40–2, 112, 113
WorldWatch Institute 103

yachts 67, 68, 71, 75–6
Yemen 68, 75, 76